1 MONTH OF
FREE
READING

at

www.ForgottenBooks.com

By purchasing this book you are eligible for one month membership to ForgottenBooks.com, giving you unlimited access to our entire collection of over 700,000 titles via our web site and mobile apps.

To claim your free month visit:

www.forgottenbooks.com/free755535

ISBN 978-0-483-10834-9
PIBN 10755535

This book is a reproduction of an important historical work. Forgotten Books uses
state-of-the-art technology to digitally reconstruct the work, preserving the original format
whilst repairing imperfections present in the aged copy. In rare cases, an imperfection in
the original, such as a blemish or missing page, may be replicated in our edition. We do,
however, repair the vast majority of imperfections successfully; any imperfections that
remain are intentionally left to preserve the state of such historical works.

PRACTICAL EXERCISES

IN

PHYSIOLOGICAL OPTICS

BY

GEORGE J. BURCH

M.A., D.Sc. Oxon., F.R.S.

OXFORD

AT THE CLARENDON PRESS

1912

HENRY FROWDE, M.A.

PUBLISHER TO THE UNIVERSITY OF OXFORD

LONDON, EDINBURGH, NEW YORK

TORONTO AND MELBOURNE

PREFACE

THIS book was written for the practical classes in Physiological Optics required by the regulations for the Diploma in Ophthalmology of the University of Oxford. I have conducted such classes for the last two years in the Physiological Laboratory, Oxford, for Professor Gotch. Hence the occasionally minute instructions with regard to particular instruments. But for the most part the descriptions are general, and it is hoped that the book may be found useful in other laboratories.

I have omitted experiments which can be shown quite well during a lecture, and have included those from which more can be learnt by the actual doing of them.

The difficulty of conducting almost any practical class is enormously increased unless the same experiment can be done by all at the same time. Unfortunately it is not possible, in any ordinary laboratory, to arrange this with a subject like physiological optics. Much of the work has to be conducted in a dark room. Some of the more expensive apparatus serves for several experiments, and very little of it is of any use in the other departments of physiology. It is therefore necessary so to organize the course as to make it possible for a number of different experiments to go on at the same time.

The instructions here given are intended to make the student to a great extent independent of frequent reference to the demonstrator.

The experiments are not all of the same value, or difficulty. Some take up a whole morning—others a few minutes only. Those marked with an asterisk are important because they involve measurement as well as observation. They should be performed by all, and a copy of the record of each student preserved in the laboratory collection.

The exact order in which the experiments are taken must depend upon the available apparatus and accommodation. Eight of them, namely IV. 1, 2, 7, 11, 12, V. 7, 8, and VI. 4, require a dark room. For fifteen others, viz. II. 1, 2, 3, 4, IV. 4, 8, 9, 10, VI. 7, 8, 9, 10, 11, 12, 13, the room must be at least partially darkened. A high-power spectroscope is required for V. 1, 3, 4, and perhaps V. 8 and VI. 7, and a low-power spectroscope for IV. 3, V. 5, and perhaps VI. 7, 8, 9.

Three observers, or better still four, are needed for IV. 1 and IV. 2, but VI. 1, 2, 3, 10, 12 can be well seen by half a dozen at once, and VI. 13 can be demonstrated to the entire class. In all other cases two should work together.

In order to prevent clashing a plan should be drawn up at the beginning of term for each of the principal instruments, and for each dark room, showing who is to have the use of it on each occasion when the class meets. A column is provided to the right of the Table of Contents in which each student can enter the dates that concern his own work. And to the left of the same Table is another column in which he should enter the date of each experiment when he has performed it. Then if he should find himself with an hour or half an hour to spare, it is easy for the demonstrator to see at a glance what he has done, and to suggest something that will occupy the time.

GEORGE J. BURCH.

Physiological Laboratory, Oxford,
April 1912.

CONTENTS

SECTION I

DIOPTRICS

SECTION II
DIOPTRICS OF THE EYE

SECTION III
JUDGEMENTS OF THE EYE: SPACE

SECTION IV
SENSATIONS OF THE EYE

SECTION VI

EXPERIMENTS BY FLASHING LIGHT

N.B.—Some of these exercises are essentially phenomena to be observed, others relate to sensations which can be measured and the results recorded. The latter are distinguished by an asterisk. Most of them require the co-operation of two or more persons each of whom should take turn in observing in order that individual differences of sensation may be recognized.

SECTION I

DIOPTRICS

Definitions.

The *Optic Axis* of a lens is the line passing through the centres of curvature of its two surfaces—or if one of them is flat, the line at right angles to that surface which passes through the centre of curvature of the other.

The *Centre of a Thin Lens* is for our present purpose practically the point midway between the two surfaces on the optic axis.

The *Centre of Figure* is the point which bisects all straight lines drawn through it to the periphery.

In a *Decentred Lens* the centre of the lens and the centre of figure do not coincide.

The *First Principal Focus* is the point on the optic axis in front of a convex lens from which if rays proceed they will become parallel after traversing the lens.

The *Second Principal Focus* is the point on the optic axis behind a convex lens to which parallel rays will converge after traversing the lens.

With concave lenses the First Principal Focus is in front of the lens and the Second Principal Focus is behind it, and both are virtual. For since such lenses cause the rays to diverge they do not come to any real focus, but behave as though they proceeded from an imaginary focus which may be found by tracing the ray backwards along the path into which it has been re-fracted by passing through the glass, for the second, or from which it has been refracted, for the first principal focus.

The *Principal Points* are two points on the optic axis such that an object on one of them gives an image on the other, of the same size, and the same way up. In double convex lenses, of equal curvature, these points coincide at the centre of the lens. In meniscus and crossed lenses, and thick lenses generally, they do not coincide.

The *Nodal Points* are such that any ray passing through the first of them will after refraction proceed as if it came from the second, in a direction parallel to that which it originally had.

Whenever, as in thin lenses and the majority of lens systems, the image is formed in the same medium as the object, the nodal points coincide with the principal points, but when as in the eye or in an immersion objective the media are not the same, then the nodal points are different from the principal points.

Focal Length is the distance from a Principal Focus to the corresponding Principal Point.

N.B.—Beyond the two principal foci is a pair of points called the Negative Principal Points such that an object on one gives an image on the other of the same size, *but inverted*. The distance from a negative principal point to the nearest principal focus is also equal to the focal length. Hence from one negative principal point to the other is four times the focal length.

This is the shortest distance at which a lens can form a real image of any object. If, therefore, in the following experiments a real image cannot be found, it is probably because the object has been placed nearer to the lens than *twice its focal length.* .

The focal lengths of English lenses were formerly always expressed in inches, those of foreign lenses on the metric system, which is now generally followed by English makers also.

The *Power of a Lens,* in *Diopters,* is the reciprocal of its focal length in metres.

Thus, a lens of 2 metres focal length is 0.5 diopter, and one of 50 cms. focal length is 2 diopters. For this purpose 40 inches is reckoned as 1 metre. So that an English lens of 10 inches would be 4 diopters.

Lens formula.

Let u = distance from object to lens.

v = distance from lens to image.

f = focal length of lens.

r_1 and r_2 = radii of curvature of the first and second surfaces respectively.

μ = refractive index of lens.

Then
$$\frac{1}{u} + \frac{1}{v} = \frac{1}{f} = (\mu - 1)\left(\frac{1}{r_1} + \frac{1}{r_2}\right), \tag{1}$$

whence
$$f = \frac{uv}{u+v} = \left(\frac{1}{\mu-1}\right)\left(\frac{r_1 r_2}{r_1 + r_2}\right).$$

Among ophthalmologists and spectacle-makers convex lenses are reckoned positive and concave lenses negative. And the formula connecting object-distance $(= u)$ and image-distance $(= v)$ with focal length $(= f)$

is written $\frac{1}{u} + \frac{1}{v} = \frac{1}{f}$, whereas in many of the books

read for examination purposes it is given as $\frac{1}{v} - \frac{1}{u} = \frac{1}{f}$.

The reason is that owing to the convention adopted in measuring these distances certain of them come out negative. Making allowance for this, u, v, and f *have*

all the same sign, and the expression $\frac{1}{u} + \frac{1}{v} = \frac{1}{f}$ is

arithmetically true.

The convention easiest to remember, and leading at once to this result, is as follows :—

Imagine yourself travelling with the light-wave, and measure all distances from point to point as you come to them in the order of time.

So long as you can go forwards all distances are positive—if you must go backwards towards the light, they are negative.

When you pass from one medium to another the refractive index of the medium you are in is to be subtracted from that of the medium you are entering.

Thus, taking the complete formula—

$$\frac{1}{u} + \frac{1}{v} = (\mu - 1)\left(\frac{1}{r_1} + \frac{1}{r_2}\right) = \frac{1}{f},$$

this is really

$$\frac{1}{u} + \frac{1}{v} = \frac{\mu_2 - \mu_1}{+ r_1} + \frac{\mu_1 - \mu_2}{- r_2} = \frac{1}{f},$$

where μ_1 = refractive index of air (= 1) and μ_2 = refractive index of glass.

The first distance, from object to lens = u, is positive. If the glass is convex, the radius of curvature, r_1, from circumference (where the light-wave is) to centre is positive because it lies farther on. But if the glass is concave, the radius r_1 is negative because you have to go back along the ray to get to the centre.

The second surface, if convex, has a negative radius, for its centre lies behind you—but as you are now passing from glass to air the larger number has to be subtracted from the smaller, and the numerator of the fraction becomes negative as well as the denominator, making the result arithmetically positive.

Those who have become accustomed to the other convention will prefer to keep to it; this, which is the older of the two, is specially convenient when using the system of diopters and dealing with complex lens systems. Moreover, the results it gives are in the form that workmen understand.

Observe that all the terms of the above formula are reciprocals, either of quantities measured or quantities sought. Now, although to add fractions with different denominators is tedious, it is quite simple if those fractions are expressed as decimals. Accordingly, in calculating by diopters it is only necessary to find the diopter value of each distance measured, by means of a table of reciprocals, and the above calculation is reduced to a matter of simple addition.

It is convenient to let diopter values be expressed by capital letters, thus $\frac{1}{u} = U$, $\frac{1}{v} = V$, $\frac{1}{f} = F$, &c.

Then (1) becomes $U + V = F = (\mu - 1)(R_1 + R_2)$ where $F = \frac{1}{f} = $ the power of the lens in diopters.

I. 1. Practical Examination of a Lens.

Holding the lens near the eye—say 1 cm. away—look through it at a fairly distant object and move the lens up and down, or from side to side, keeping it always approximately at right angles to the line from the object to your eye.

If the object appears to move in the *opposite direction* to that in which you are moving the lens, the lens is *convex*.

If the object appears to move in the *same direction* as that in which you are moving the lens, the lens is *concave*.

If the object *does not move* either way, the lens is a *plane parallel plate*.

If, when the lens is interposed, the object appears *displaced towards one side of it*, but does not move when the lens is moved from side to side or up and down, the lens is a *prism with plane faces*. In this case if the lens is rotated about the optic axis the object will appear to describe a small circle.

Cylindrical Lenses behave as plane parallel plates in the direction parallel to the axis of the generating cylinder, and as lenses in the direction at right angles to it. Consequently a direction can be found in which if the lens be moved to and fro the object will appear to remain stationary. This coincides with the axis of the cylinder. At right angles to this direction the object will move with the lens if the cylinder is concave, or in the opposite direction if it is convex.

In most cases one surface is spherical and the other cylindrical, so that there is no direction in which the movement is zero, but with a little practice the directions of maximum and minimum movement can be recognized.

Occasionally both curvatures are found on the same surface, which is then said to be toroidal or toric. The surface of a bicycle tyre is toroidal, the radius of curvature being in one direction half the diameter of the wheel and in the other half the diameter of the tyre.

Further Examination of Cylindrical, Spherocylindrical, Toroidal, and Decentred Lenses.

Rule upon a card two straight lines at right angles to each other. Fix the card at a convenient distance—2 or 3 feet from the eye will serve in most cases, but it may be less for strong lenses and more for weak ones.

Hold the lens midway between the eye and the card. It the centre of the cross as seen through the lens falls on the optic axis, the four arms of it will coincide with the four arms visible outside the lens. If this takes place when the centre of the cross is at the centre of figure, the lens is truly centred. But if, to make the four arms fit, it is necessary to bring the image near the edge of the lens, then it is decentred.

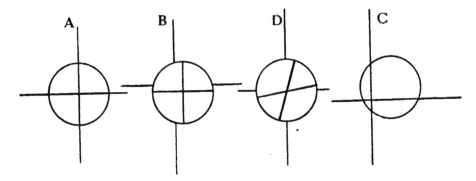

FIG. 1.

In Fig. 1 A shows a truly central lens ; B indicates decentring, and c shows the centre of the cross marking the position of the optic axis, which is obviously decentred.

If the lens is cylindrical, sphero-cylindrical, or toroidal, on rotating it about the optic axis the arms of the cross will not continue at right angles to each other. For a cylindrical lens does not magnify in the direction parallel to the axis of the cylinder, but does magnify— or diminish—at right angles to it. Hence we may have for one position of the cylinder, as in D, the upper right and lower left quadrants unaltered, while the upper left and lower right quadrants are magnified, with the result that the lines are no longer at right angles. This, there- fore, is the characteristic test for such surfaces.

Moreover, if we continue to rotate the lens, the obliquity of the lines will reach a maximum and then diminish, and they will appear exactly at right angles when the cylindrical axis coincides with either of them.

This, therefore, is the test for the direction of the axes.

If the lens is a simple plano-cylinder, then moving it from side to side while holding it close to the eye will show which of the two lines is parallel to the cylinder axis, because in that direction it will behave as plane glass and the image will remain still.

If the lens is a sphero-cylinder, there will be a direc- tion of minimum movement and one of maximum at right angles to it. It remains to ascertain which of these is parallel to the axis of the cylinder.

Breathe on one side of the lens to dull it, and try whether it gives an image by reflection from the other— an incandescent electric lamp affords an excellent object for this purpose. If the spherical surface is concave, you will get a true image of the lamp. In this case, whatever convergent effect there is must be due to the second surface, and if this is cylindrical, on applying the test of movement, the image would go *with the lens* when moved parallel to the cylindrical axis, and *against the lens* at right angles to it ; and vice versa if it is concave.

If the cylindrical surface is concave it will give by reflection a 'line-image' of the lamp, parallel to the axis of the cylinder.

If both surfaces are convex, the phenomenon illustrated in Fig. 1 D will still give the axis which must lie in the quadrants of greatest magnification.

From the foregoing it will be seen that by the three tests of moving the lens in front of the eye, the crossed lines, and the reflected images, the general character of a lens may be diagnosed.

There are three combinations, viz. a toroidal with a convex sphere, with a convex cylinder, and with another toroidal, that cannot be recognized with certainty by these tests.

I. 2. To find the Focal Length and Power of a Convex Lens.

Apparatus required :—One lens-holder, two adjustable cardboard pointers with plumb-lines on stands, a black screen, and a metre rule, or better, two.

N.B.—In order that there may be a real image, the screen must be *at least four times the focal length of the lens* from the object.

The method here described requires a minimum of apparatus, measurements can be made to within 1 millimetre—a degree of accuracy amply sufficient—and with a little practice the operation can be completed in five minutes.

1. Without some arrangement for cutting off extraneous light it would be impossible to identify the image of the object formed by the lens. A black screen is therefore placed at one end of the table behind the white cardboard pointer which constitutes the object. This screen should be 2 feet square or more.

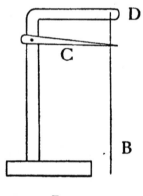

FIG. 2.

The pointer C (Fig. 2) is a strip of card cut to a fine point, attached by a drawing-pin to the upright stand A, from the overhanging arm of which is suspended a plummet B by a thread passing over a pin D. The pointer must be set at right angles to the line of sight and should point towards the edge of the bench. It is adjusted vertically by rotating it about the pin, and horizontally by moving the stand.

The lens-holder is large and black, so as to act with the screen as a background against which to see the image (Fig. 3). It consists of a thin board supported vertically at right angles to the edge of the table on three feet, and furnished with two stops working against the edge of the table so as to guide it in a straight line.

2. Fix the lens over the hole in the holder by means of the three springs. Place the holder some distance

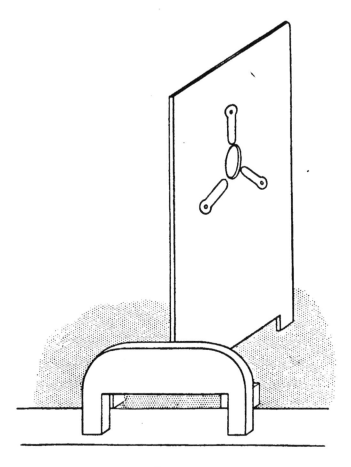

FIG. 3.

from the object, i. e. the cardboard pointer, which must be adjusted to the level of the centre of the lens.

3. Going close to the lens, look through it at the tip of the pointer—which will be visible, though blurred and indistinct—then move back along the line of the

bench, keeping the pointer in sight. It will grow less distinct but larger, until quite suddenly it expands into a blur of white that fills the lens. Keep on moving farther away in the same straight line for some 30 cms. and you will see the blur of white contract again until it resolves itself into a sharply defined image of the pointer, reversed, i. e. pointing *away from* the edge of the bench.

Now half the art of measuring focal lengths quickly has to do with the finding of this image, and the average student spends an amount of time over it that is perfectly absurd.

It should be done with a single uninterrupted movement, and unless the lens is of very long focus the image should be found with certainty each time. The best way is to practise with a lens of short focus—say 15 cms.—and blacken all of the cardboard pointer except an inch of the tip. Place the lens first 60 cms. from the pointer and practise moving the eye back, keeping the white blur of the pointer in sight till it expands and fills the lens and finally contracts to a clear image. Place the other pointer so that it appears to touch the image, point to point. You will then find that the reason why at a certain distance the white blur expands to fill the lens, is because when in that position the *image is focused on the pupil of your eye*. It becomes clear and sharp as soon as you are far enough away to see it distinctly. But it is only visible when your eye is in a line with the object and the centre of the lens. If therefore the image is too far to the left, the object is too far to the right—if the image is too high above the table the object is too low and must be raised.

Three trials should suffice for the adjustment of the

object, so that the line of sight may be approximately parallel with the edge of the table.

4. To adjust the second pointer accurately to the image of the first, move your head up and down whilst looking at them. They will both appear to move relatively to the lens, but the one that moves fastest, i. e. most, is nearer to your eye than the other. If this is the second pointer, push it farther away—if the image moves fastest, bring the second pointer nearer to you, adjusting its height and position so that its point appears to touch the point of the image. When you have done this as accurately as you can with the naked eye, examine it with a magnifying lens of about 2 inches focal length. You will now see a more or less clear image of the first pointer near the tip of the second. Move the second pointer nearer or farther until both are sharply focused at the same time, and their points appear in contact and remain in contact while you move your eye.

N.B.—The image will probably not be visible through the lens at all unless you get the adjustment very nearly right with the naked eye first.

5. Everything is now ready for measurement. Since the line of sight, or optic axis, is parallel to the table, the distance from object to lens is given by the distance of the first plummet from the foot of the lens stand, and is read off on a metre rule lying on the table under the plummet. The distance from lens to image is read off in a similar manner, the lens stand being set vertically and being cut away underneath so that the metre rules will pass under it, to enable both measurements to be taken to the same face.

The correction for thickness can be added to one and subtracted from the other afterwards.

Then $\dfrac{1}{u} + \dfrac{1}{v} = \dfrac{1}{f}$ in cms.,

or $\dfrac{100}{u} + \dfrac{100}{v} = D$ in diopters = power,

or $\dfrac{uv}{u+v} = f$ in cms. = focal length.

Having completed the first measurements, set the object 40 cms. from the lens instead of 60.

It is very helpful both to calculate the results and to measure them.

From the formula $v = \dfrac{fu}{u-f}$

u	$\dfrac{fu}{u-f}$	v	$u+v$
60	$\dfrac{15 \times 60}{60-15}$	20	80
40	$\dfrac{15 \times 40}{40-15}$	24	64
30	$\dfrac{15 \times 30}{30-15}$	30	60
20	$\dfrac{15 \times 20}{20-15}$	60	80
10	$\dfrac{15 \times 10}{10-15}$	-30	

It will be observed that by bringing the object nearer to the lens by equal stages the image is moved farther away by an amount which continually increases, becoming infinite when the object is at the first principal focus

$$\left(\text{if } u = 15, \quad v = \frac{15 \times 15}{15-15} = \frac{1}{0} \right).$$

But if u is smaller than f, v becomes negative, that is to say, there is no real image, but only a virtual one. In nine cases out of ten when a beginner is unable to find the image it is because he has made u smaller than f, in other words has *placed the object too near the lens.*

Observe further that the sum total of $u + v$ is smallest when $u = v$, and is then equal to $4f$. The practical meaning of this is that the lens should always be placed to begin with at the middle of the bench and the object at one end of it. If the image of the pointer does not expand to fill the lens within the length of the bench it means that the focal length of the lens is greater than can be dealt with in this manner. Spectacle lenses are made of $1 D$, $0.75 D$, $0.5 D$, and $0.25 D$. From what has been said a lens of $1 D$, having a focal length of 1 metre, would require a bench 4 metres long. A lens of $0.5 D$, focal length 2 metres, could not give a real image less than 8 metres from the object, and one of $0.25 D$ would require a bench 16 metres long. One mode of dealing with such lenses is to place object-pointer, lens, and image-pointer on three separate small tables and measure the distances between the table-tops by hanging plumb-lines from them to the floor.

Lenses of Great Focal Length.

But there is another method as follows: Place the object-pointer a good distance in front of the lens, and put a plane mirror behind it. Find the position of the image with the eye, as already described, and mark it by a second pointer. Shift the object-pointer a little nearer the image-pointer and find the new position of the image. Repeat until object-pointer and image coincide. The object is then at the principal focus of the lens. The rays falling on the mirror and consequently those reflected back by it, are parallel, and they form an image at the second principal focus of the lens, which in this case coincides with the first. Accordingly the distance from the pointer to the centre of the lens is the focal length.

I. 3. Focal Length of a + Cylindrical Lens.

Apparatus required:—Black screen, lens-holder, cardboard pointer with plumb-line, and special 'object' consisting of a disk with cross-wires at right angles mounted in a stand in which it can be rotated.

Theory of the Method:—Rays proceeding from a point after passing through a·+ cylindrical lens converge to a line parallel to the axis of the cylinder at a distance v from it, such that

$$\frac{1}{u} + \frac{1}{v} = \frac{1}{f},$$

where u and f have the same meaning as for ordinary lenses.

Hence, using as object a line parallel to the cylinder axis, a bright and well-defined linear 'image' is formed, which though not a true image will serve the purpose of one. Using two lines at right angles affords a ready means of ascertaining when one of them exactly coincides with the cylindrical axis by the test given in exercise I. 1 and Fig. 1 D.

1. Set up the disk with the cross-wires in front of the black screen—fix the lens in the holder, and place

it not less than twice the supposed focal length from the cross-wires.

2. Observe the cross-wires through the lens and rotate the disk in the holder until the two wires appear at right angles to each other, and one of them looks sharp and the other probably blurred.

N.B.—With a simple cylindrical lens, i.e. one of which neither surface is spherical, the wire at right angles to the cylindrical axis will appear sharp at all distances.

Fixing the attention on the wire that looks blurred, move the eye to a greater distance. The wire will seem to thicken until it fills the lens. If in so doing it *keeps at right angles to the other the disk is correctly set*, if not, adjust the lens by rotating it very slightly in the holder. Then move the eye a foot or so farther away and you will see a sharply defined image of the wire parallel to the optic axis and therefore at right angles to the other which has been visible all the time.

Mark the position of it with a cardboard pointer, adjusting by parallax, first with the eye and finally with the aid of a lens. But observe that you will only see one wire with the lens, though with the naked eye you will see both, because the line image at right angles to the cylinder axis coincides with the object, whereas that parallel to the cylinder axis is close to the pointer.

The measurements are exactly the same as those for an ordinary lens, and so are the calculations

$$\frac{1}{u} + \frac{1}{v} = \frac{1}{f} = D,$$

$$f = \frac{uv}{u+v}.$$

If the second wire does not expand to fill the lens the object is too close to the lens.

It is well to measure two or three of different focal lengths before proceeding to the more difficult sphero-cylinders and toroidals.

I. 4. To determine the Powers of + Sphero-cylinders and + Toroidals.

Apparatus required :—The same as for simple cylindrical lenses.

The mode of procedure is the same as for simple cylindrical lenses, but the preliminary setting of the wires can only be done when the lens is at a certain distance, easily found, from them.

1. Place the eye close to the lens and try various distances, rotating the disk until one of the wires is well defined. Then move both the eye and also the disk farther from the lens—if both wires, though blurred, still appear at right angles to each other the disk is correctly adjusted, if not, it must be cautiously turned so as to bring them so.

2. If the sphero-cylinder, or toroidal, is positive in both directions (see I. 1) and is far enough from the lens, you will find a sharp image of one wire at a certain distance v_1 from the lens, and an equally sharp but larger image of the other at right angles to it at a greater distance v_2. Since the distance u of the object is the same for both we have

$$\frac{1}{u} + \frac{1}{v_1} = \frac{1}{f_1} = D_1,$$

$$\frac{1}{u} + \frac{1}{v_2} = \frac{1}{f_2} = D_2$$

where D_1 and D_2 are measured in diopters and represent the 'power' of the lens in two directions, namely parallel and at right angles to the cylindrical axis. The question is, which is which?

Suppose both sphere and cylinder are positive. Then the weaker power D_2, corresponding to the greater distance v_2, must be that due to the spherical surface.

And since this exerts an equal power in all directions it follows that the difference $D_1 - D_2 = C$ is the 'power' in diopters of the convex cylindrical surface.

But the same effect so far as this formula is concerned would be produced by a stronger sphere with a concave cylinder, and we should then have

$$D_2 - D_1 = -C,$$

the power in diopters of the concave cylindrical surface.

3. If either surface is toroidal the method of reflections may be used to obtain a complete determination of the lens.

I. 5. To find the Focal Length and Power of a Concave Lens.

Apparatus required:—The same as for convex lenses with the addition of a convex lens in an adjustable stand, and a third cardboard pointer.

To any one who can depend on not displacing the apparatus he has already adjusted, the determination of a concave lens offers no difficulties.

1. One of the cardboard pointers is set up near the end of the table and the convex lens placed so as to throw an image of it at a good distance from itself.

The position of this image is accurately found and marked by the second pointer. The concave lens is then placed in the path of the rays between the convex lens and the second pointer. Put the concave lens at first fairly near the pointer and note that it makes comparatively little difference in the position of the image. But on moving the concave lens nearer to the convex lens—i. e. *against* the direction of the light-wave—the final image moves farther away from it and becomes larger. Observe that this apparent motion of object and image in opposite directions is merely accidental. In all lenses or lens systems, if the object is moved along the axis, its image moves *in the same direction.* In this case the 'object' is what is called 'virtual', i. e. it is only formed if the concave lens is not there. Moving the concave lens back comes to the same thing as moving the virtual object forward—and as has been stated, the final image also moves forward.

Let the distance from the virtual object, as marked by the second pointer, to the concave lens be $-u$. This is negative, being measured against the light. Let the distance from the concave lens to the final image, of which the position must be ascertained in the usual way and marked by a third pointer, be $+v$ (measured *with* the light).

Then $\dfrac{1}{-u} + \dfrac{1}{+v} = \dfrac{1}{f}$, and as v is greater than u, f will be negative.

The most accurate results are obtained if the distances are selected so as to make v about twice as great as u.

If you have a convex lens of known power, stronger than the concave, place the two in contact in the lens-holder and determine the power of the combination, treating it as a single convex lens.

Let $D_1 =$ the power of the convex lens,

$D_3 =$ the power of the two combined.

Then $D_3 - D_1 = D_2 =$ the power of the concave lens which will have a negative value.

This method is not quite so accurate as the previous one, but is near enough for practical purposes in dealing with thin lenses.

I. 6. Focal Length of a −Cylindrical Lens or of a −Sphero-cylinder or −Toroidal.

This may be measured quite easily by substituting the disk with cross-wires for the first cardboard pointer in the method for concave lenses (I. 5) and using the virtual object so formed.

It is a little difficult at first to find the correct position for the toroidal. As a rule it should be such that the two cylindrical images are not too much magnified.

But the value of the experiment is not so much its practical utility as the insight it gives into the properties of these lenses and the position along the optic axes of the line-images formed by them.

I. 7. To measure the Power of a Prism in Prism Diopters.

Apparatus required:—A small lens-holder, five stands with plumb-lines, metre-rule.

1. Fix the prism in the lens-holder so that the direction of its refraction may be horizontal. Any

simple holder will serve, the large black one used for focal lengths being unnecessary. Place two plumb-lines behind it, one at a distance of 5 or 10 cms. and the other 40 or 50 cms. farther away. Keeping your eye 40 or 50 cms. in front of the prism in such a position that the two plumb-lines appear one behind the other, place a third in line with them 5 or 10 cms. in front of the prism, and a fourth 30 or 40 cms. nearer the eye.

2. It is quite easy, by going a little farther off, to see when the four plumb-lines appear exactly in a line. They must be accurately adjusted to do so, and the prism must be so set that the refraction is equally divided between its two faces. This is easily tested by turning it backwards and forwards about a vertical axis. The apparent displacement of the object is least when both faces make equal angles with the optic axis. This is called the position of minimum deviation. It remains to measure the angle between the original incident and the final refracted rays.

3. Take the prism and its stand away and place the fifth plumb-line 40 or 50 cms. behind where it stood so as to be really in a line with those in front. The angle between the apparent and the real position of the plumb-lines seen through the prism expressed in 'prism diopters' is its power.

Pin two sheets of paper to the table, one to include the intersection of the lines through 4, 3, 5 and 2, 1 produced, and the other so that it may contain the same lines produced, at a distance of either 50 cms. or 100 cms. from their intersection.

(N.B.—Three flat-bottomed lead weights of a pound

each will hold down the paper in a quite reliable way if pins cannot be used.)

The distance of the lines apart at 1 metre from their intersection, measured in centimetres, on a scale at right angles to one of.the lines, is the power of the prism in 'prism diopters'. From it the 'angle of deviation' can be found by a table of tangents.

The relation of the refracting angle of the prism, i.e. the angle between its faces and the power in diopters, depends on the refractive index of the glass. For thin prisms such as are used by ophthalmologists the refracting angle in degrees is usually about 10 per cent. greater than the power in diopters.

I. 8. The Lens-gauge and the Spherometer.
[Figs. 4 and 5.]

For technical purposes rapid measurements of the power of lenses are effected with the lens-gauge.

This instrument, which in size and shape resembles a watch, has three metal pins in a row, projecting from its side. Of these, the two outer are fixed, and the middle one, which normally projects beyond the other two, can be depressed below their level, the amount of its displacement being indicated on the dial by the movements of a hand, like that of a watch, with which it is connected.

The theory of the instrument is based on Euclid III. 35. 'If two chords of a circle cut one another at a point within the circle, the product of the segments of one chord is equal to the product of the segments of the other chord.'

Suppose the lens-gauge is applied to the surface of a sphere (Fig. 4).

The two outer pins touch it at P and P_1 and the middle pin at A.

Draw AB, a diameter of the circle, of which the centre is C. Join PC and P_1C.

Draw PP_1 cutting AB in T.

Then PP_1 is the distance between the points of the

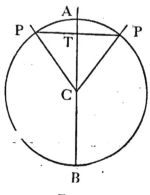

Fig. 4.

outer pins, and is constant. Also AT is the (variable) distance of middle pin above or below the line of the other two, and $PT = P_1T$.

Since C is the centre of the circle $PC = AC = BC$, but $AT = PC - TC$ and $TB = PC + TC$,

$$\therefore \ AT \times TB = PC^2 - TC^2.$$

And by Euclid I. 47 $PT^2 = PC^2 - TC^2$,

$$\therefore \ AT \times TB = PT \times P_1T.$$

Let $PT = r$, and let $AT = h$, and $2R$ = diameter of the circle.

Then

$$h(2R - h) = r^2,$$

$$R = \frac{r^2}{2h} + \frac{h}{2}.$$

If the lens-gauge is not held upright but is tilted in either direction on the surface, its readings are increased. Hence, for accurate work, a spherometer is used, in which the support consists of three fixed legs in an equilateral triangle, and the movable fourth leg is situated at the centre of it.

The calculation is the same, but as the measurements are the distances d from leg to leg instead of their distance r from the central leg, the data are different.

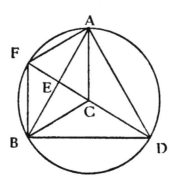

FIG. 5.

Let ABD (Fig. 5) be the equilateral triangle inscribed in a circle of centre C. Produce DC through E to F on the circle.

Then $FE = EC = \frac{1}{2}r$ and the angles ECA and ECB are equal.

Hence by Euclid I. 47 $AE = \sqrt{1 - \frac{1}{4}} = \frac{1}{2}\sqrt{3}$ and $AB = d = \sqrt{3}$.

$$\therefore r = \frac{d}{\sqrt{3}}.$$

Substituting $\qquad R = \dfrac{d^2}{6h} + \dfrac{h}{2}.$

C 2

The curvature of a lens is connected with its focal length by the formula

$$\frac{1}{f} = (\mu - 1)\left(\frac{1}{r_1} + \frac{1}{r_2}\right),$$

where r_1 and r_2 are the radii of curvature of the surfaces.

Or in diopters $F = (\mu - 1)(R_1 + R_2)$.

It is apparent, therefore, that the focal length varies with the kind of glass used.

A lens-gauge measures the *curvature of each side separately*, but it is usually graduated to read the *power* of each side in diopters, on the supposition that the index of refraction of the glass is 1·51. But this value is not always taken. For accurate work therefore, and always in measuring the power of mirrors, it is necessary to know for what refractive index the gauge is designed.

To ascertain this, measure the radius of curvature of a concave lens by the method of reflection and compare it with the results obtained with the lens-gauge.

Or still better, compare the lens-gauge with a spherometer.

I. 9. Graphic Methods.

Both for academic purposes and in practical work it is frequently necessary to trace graphically the course of rays through a lens or lens system.

Three rays are in general available from any point of the object not on the optic axis, as follows :—

1. The ray through the first principal focus becomes parallel to the axis after passing the lens.

2. The ray parallel to the optic axis passes through the second principal focus after passing the lens.

3 a. The ray through the centre of the lens proceeds on its course without deviation. (True only of thin lenses.)

3 b. The ray through the *first nodal point* proceeds after refraction as if it came. from the *second nodal point*, in a direction parallel to that it originally had. (True of all lenses.)

Any two of these rays being traced from a point on the object, their intersection gives the corresponding point on the image.

In general the two that intersect at the greatest angle should be taken.

But the formula, which this construction represents, is only true when the angles of incidence and refraction are so small that their sines and tangents are sensibly equal. Thus a thin lens 3 cms. in diameter and of 100 cms. focal length gives an image that will bear magnifying. On the other hand to trace the rays graphically through a lens of these dimensions would lead to hopeless inaccuracies.

Accordingly in finding a focal length *practically* the lens should be stopped down, especially if it is a thick one, to $\frac{1}{30}$th or $\frac{1}{40}$th of its focal length.

But in finding the position of an image, &c., *graphically*, remember that the *construction is true for any angle*, and that the accuracy attainable is greatest with a fairly large one.

1. *Given the Focal Length of the Lens and the Position of the Object, to find that of the Image.*

For instance, suppose the problem set is to find graphically the position and size of the image of an object 4 cms. high at a distance of 30 cms. from a lens of 20 cms. focal length and 4 cms. in diameter. By calculation it is easy to see that the image is formed at 60 cms. from the lens on the other side, making a total distance of 90 cms. between object and image. To make a diagram of this on foolscap paper it must be reduced to $\frac{1}{4}$, i. e. $22\frac{1}{2}$ cms. The object would then be only 1 cm. high.

A ray parallel to the optic axis just passes through the lens, and may be traced onwards through the second principal focus. But the ray from the same point through the first principal focus appears in the diagram to miss the lens altogether, passing by it at too great a distance from the axis. Hence, beginners often think it necessary to use the ray through the centre of the lens, and this makes so small an angle with the other that it requires some care to get the position of their intersection reasonably correct.

The best plan is to regard the construction simply as a graphic method of finding the position of the image-plane corresponding to a given object-plane and focal length.

Thus in the example given :—

Make the diagram $\frac{1}{4}$ full size.

Draw an optic axis, with an object-plane and a lens-plane $7\frac{1}{2}$ cms. apart, at right angles to it. Mark the first and the second principal focus at distances of 5 cms. from the lens-plane on the optic axis.

Take any convenient point on the object-plane as far away from the optic axis as possible, draw a line from it through the first principal focus to the plane of the lens and continue it parallel to the optic axis. Draw a second line parallel to the optic axis as far as the lens-plane and continue it thence through the second principal focus. The intersection of these lines is the 'conjugate point' on the image-plane.

A second point may be taken on the object-plane— preferably on the other side of the optic axis and as far as possible from it, and the process repeated. The 'image-plane' is determined by drawing a line through the two conjugate points thus found. The actual course

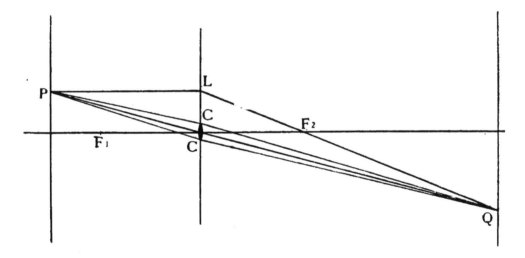

Fig. 6.

of the rays from the object given through the lens may now be represented.

Draw the object to scale on the object-plane, and the lens to scale on the lens-plane (Fig. 6). Then with a straight-edge rule a line from the top of the object through the centre of the lens on to the image-plane, and another similarly from the bottom of the object. These will give the size of the image, and two lines from each point, one to the top edge of the lens and the other to the bottom of it, will indicate the cones of rays forming it. It will be seen that these lines sub-tend far too small an angle to serve for the graphical construction.

2. *Given the Positions of Object, Lens, and Image, to find the Focal Length by a Graphic Method.*

Draw an object-plane, lens-plane, and image-plane in position on an optic axis (Fig. 7).

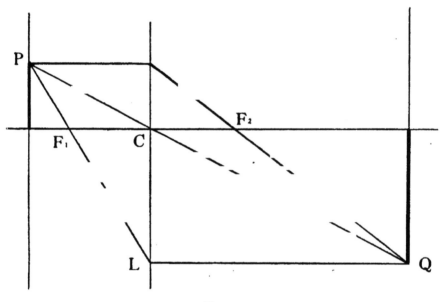

FIG. 7.

From any point P on the object-plane a good distance from the axis, draw the ray through the centre of the lens to the image-plane. From the conjugate Q thus found, draw a line parallel to the optic axis cutting the lens-plane in L. Join PL. The point F_1, where this cuts the optic axis, is the first principal focus and F_1C is the focal length.

F_2 may be found in a similar way.

3. *Given the Distance between an Object and its Image and the Magnification required, to determine the Position and Focal Length of the Lens.*

Draw an object-plane and an image-plane the specified distance apart on an optic axis (Fig. 7).

Measure from the optic axis on the object-plane a length OP and from the optic axis on the image-plane, in the opposite direction, a length IQ, such that

$$\frac{IQ}{OP} = M = \text{the magnification required.}$$

Join PQ. The point L, where PQ cuts the optic axis, is the position of the lens-plane.

The focal length of the lens can be found as in paragraph 2.

I. 10. Properties of Thick Lenses.

The properties of thick lenses may be studied by means of a glass tank 10 inches by 6 inches by 3 inches deep, and two thin lenses, one plano-convex and the other plano-concave, of 3 inches focal length.

The lenses must be mounted in such a manner that their flat faces can be applied to the ends of the glass tank, which is filled with water.

For most purposes a thick lens may be regarded as two thin lenses separated by a plate. In this case the 'plate' is of water and the 'lenses' of glass, but as the sole function of the lens portion is to produce convergence or divergence of the rays, the result is the same as if the lens were of water, only in that case the curvature would need to be stronger because of the lower refractive index of the medium.

1. Set up an object some distance in front of the curved side of the plano-convex lens, and find the position of its image in the usual way.

Apply the formula $\dfrac{1}{u} + \dfrac{1}{v} = \dfrac{1}{f}$ and calculate f.

Then, without moving the lens, bring up the tank in contact with its flat side. The image is now formed in the water. Let a plumb-line from one of the stands hang in the water, and use it to determine the position of the image in the ordinary way by parallax. The new distance v_1 will be longer than the old in the proportion of the refractive index of water to that of air.

But the focal length of the lens-part being unaltered, in order to make the calculation give the same value for f, we must write $\dfrac{1}{u} + \dfrac{\mu}{v_1} = \dfrac{1}{f}$ where μ is the refractive index of water.

2. Find the position of the Principal Foci. If the sun is shining, reflect sunlight through an open window first through the lens, receiving the image in the water, and then the other way through the water, receiving the image in the air. Measure the distance in each case from the vertex of the curved surface.

If the sun is not available use the method for lenses of very great focal length described on p. 25.

Place a mirror at the far end of the tank and find the position in front of the lens in which an object coincides with its reflected image.

Then, reversing the arrangement, place the mirror

on the other side of the lens, and looking through the tank find the position of an object in the water so that it may coincide with its reflected image.

Measure the distances as before, to the vertex of the refracting surface. The results should show that the focal length in water is 1·3 times the focal length in air.

What we commonly call the focal length is therefore more accurately the *focal length in air*, and the general lens formula should run

$$\frac{\mu_1}{u} + \frac{\mu_2}{v} = \frac{\mu_0}{f} = \frac{1}{f'},$$

where μ_1 is the refractive index of the first medium, μ_2 that of the second, and μ_0 that of the medium in which the focal length is to be reckoned, namely, air.

The two familiar instances of this condition of things are the eye and the immersion objective of a microscope. The equivalent focus in air of the objective is quite different from the actual focus in glass and cedar oil and balsam.

Helmholtz gives as the position of the first principal focus of the eye a point 12·83 mms. in front of the cornea, and that of the second 14·65 mms. behind the back of the lens, the first focal length (i. e. in air) being 15 mms., and the second (in the eye itself) being a little more than 20 mms.

In a thick lens, of which only one refracting surface comes into action, the principal points coincide where the optic axis cuts the surface. The nodal points coincide at the centre of curvature, which is clearly the only point fulfilling the conditions, namely, that every incident ray directed towards the first nodal point will after refraction proceed as though it came from the second nodal point in a direction parallel to that it had at first.

In the eye there are practically three refracting surfaces, namely the cornea and the front and back of the lens. Consequently the two nodal points do not coincide, but are nearly four-tenths of a millimetre apart, and lie rather nearer to the cornea than its centre of curvature by about $\frac{3}{4}$ mm. and $\frac{1}{3}$ mm. respectively.

3. To find the position of the Nodal Points. Hang two plumb-lines in the water some distance apart where they can be seen magnified on looking through the lens. Place a third plumb-line in such a position that when it and either of the others appear in line on looking over the top of the tank their images also appear in line as seen through the water. This can only be the case if the third plumb-line passes through the nodal points.

SECTION II

DIOPTRICS OF THE EYE

II. 1. Skiascopy.

The theory of Skiascopy may be studied with advantage as follows :—

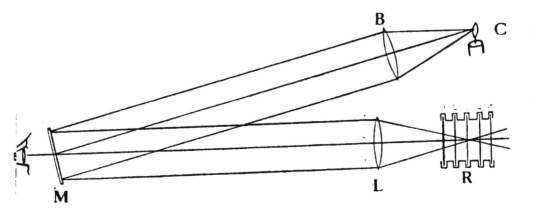

Fig. 8.

To represent the lens, &c., of the eye use a 3-inch reading-lens L of about 6 inches focus. To represent the retina, place five plates of clear glass in an ordinary plate-rack R, about an inch apart, after dusting a little French chalk or flour on each to make the course of the rays visible.

Stand the plate-rack on a box on the table so as to be about the height of the eye of a person sitting in a chair. Support the lens in one of the adjustable stands by it at such a distance that parallel rays after refraction through it may traverse all five plates, being brought to a focus on the middle one.

Behind the plate-rack and to one side of it place a lamp C in the principal focus of another lens B, so that it may give parallel rays. This lamp must be screened from the plate-rack and the reading-lens.

Then let one observer reflect light from the lamp into the reading-lens with an ophthalmoscope mirror M, while the other watches the movement of the ray in its course through the five plates.

The addition of a suitable convex lens will shift the focus on to the second plate, and a concave lens will throw it back to the fourth plate.

As by inclining the mirror M the ray reflected from it falls on different parts 1, 2, 3, of the lens L, it is refracted in the directions indicated by the lines bearing those numbers. If the eye is hypermetropic it is as though the retina were in the position of the second plate. The rays not having crossed, the 'shadow' crosses the retina in the same direction as the reflected beam moves on the lens L. A suitable convex lens placed in front of L will bring the focus forward on to the second plate.

If the eye is myopic it is as though the retina occu-pied the position of the fourth plate—the rays have crossed before they reach it and consequently the 'shadow' appears to move in the opposite direction to the beam.

If instead of a plane mirror a concave mirror is used these phenomena are reversed.

Suppose a plane mirror reflecting parallel rays is

rotated from left to right so as to make the beam travel slowly across the lens L. Then the first ray touches the lens on the left side and comes from the right side of the mirror.

But if a concave mirror N is used at such a distance that the rays come to a focus and cross before reaching the lens, then the first ray that touches the left side of L comes from the left side of N. And therefore the last ray, which comes from the right side of N, is more inclined to the axis than the first was. Hence the apparent movement of the shadow in the opposite direction.

The ophthalmoscope mirror may with advantage be supported in one of those physiological stands that provide for a small angular motion about a vertical axis. The movement of the image will be easier to follow than if it is held merely in the hand.

.After this large scale experiment practice with the artificial eye will offer fewer difficulties and be found more instructive.

II. 2. The Blue-Glass Test.

This test depends on the fact that the eye is not achromatic. Consequently in the cone of rays refracted by the cornea and the lens towards the retina, the outside rays are red, the blue being refracted at a greater angle. Beyond the apex of the cone the order is reversed, the blue coming on the outside. If, focusing the eye on a distant window bar, we bring slowly into line with it a paper-knife held in the hand, the edge of the nearer body appears bordered with red. But if we focus the paper-knife, the more distant window bar appears bordered with blue. This phenomenon is greatly intensified by using blue or purple glass, by which all the green and yellow rays are stopped and only red and blue-violet are transmitted. The object best suited for the purpose is the filament of a glow-lamp—failing this, a hole or slit in a card with a bright light behind it. If the object is nearer than the eye can focus, it will appear bordered with red ; if beyond the limit of accommodation it will be surrounded with a blue halo. When the power of accommodation is lost the distance focused with a given pair of glasses can be determined within a centimetre or so by the absence of colour round the image.

II. 3. Professor Gotch's Purple-Glass Test.

This differs from the Blue-Glass Test in two respects. The colours are more widely separated, the absorption extending farther into the blue. The coloured glass is covered with a metal plate in which is a pinhole through which the observer looks at the filament of an incandescent lamp, slowly moving the pinhole from side to side in front of the pupil of his eye. He sees two images of the filament, one red and one violet-blue. If the lamp is *within* the distance focused the *red* image appears to cross over from side to side, the violet moving comparatively little, but if it is *beyond* the focus the *violet* image swings across and the red one moves least. At the focal distance both move together. This method is more accurate than that previously described.

Those who have not lost the power of accommodation may find some difficulty at first in seeing this phenomenon. In such cases it is better to try the experiment with a strong pair of spectacles so that there may be an artificial *punctum proximum* and *punctum remotum* fairly close together.

*II. 4. The Helmholtz Ophthalmometer.

In the ophthalmometer we examine through a telescope the images of two bright objects reflected in the eye.

If the actual distance between the objects and their distance from the eye are known, and also the distance apart of the two reflected images, then it is a simple problem in geometry to determine from the laws of reflection the diameter of the sphere which would give two images with that space between them.

We may either keep the objects fixed and measure the distance between the reflected images, or separate the objects until there is a fixed distance between their images.   Helmholtz adopted the former plan.

The telescope is mounted altazimuth fashion so as to move about a vertical and a horizontal axis, and it can also be rotated about the axis of its own tube.   It is furnished with counterpoises that it may remain in any given position.

In front of the object-glass, each covering half of it, are two parallel plates of glass 4 or 5 mms. thick, each mounted on an axis at right angles to the optic axis of the telescope, and so connected that they turn in opposite directions.

Now when an object not too far off is observed through a thick plate of glass, if the plate is inclined the object is displaced sideways by an amount depending on the thickness of the plate and the angle of inclination, but *not on the distance of the object.* The reason why the movement cannot be *seen* in distant objects is simply that it is *a parallel displacement of the ray*, not an angular motion of it, and although a change of position through 2 or 3 millimetres is easily seen close to, it passes unnoticed 50 metres away.

Two plates are used because not only is there a parallel displacement of the ray, but the image is brought a little nearer through the refractive power of the plate, and unless both images were treated alike they could not be focused simultaneously.

On looking through the telescope, both plates being at zero inclination, only a single image of each object is visible. But on turning the milled head by which the plates are inclined everything is doubled, the two images moving in opposite directions.

It is the amount of this doubling that has first to be measured.

1. *Calibrating the Instrument.*

Set up, at a convenient distance—say 1 metre—from the instrument, a well-engraved millimetre scale, that of a vernier gauge being suitable. See that the scale is normal to the line of sight, and focus the telescope upon it. Set both plates to zero and note whether the image of the lines is truly single; if not, make it so by turning the milled head, and note the readings on the graduated circles. Then turn the milled head until the separation of the images is exactly 1 millimetre. This can be determined with great accuracy because the lines will exactly overlap and produce what appears to be a single image with doubling only at the centimetres and half centimetres.

The divided circles are read for this also and then the milled head turned until the images overlap by 2 millimetres, and so on.

Beyond an overlap of 7 millimetres everything goes dim, then clears again, and the overlap is found to be in the opposite direction.

By looking into the open end of the telescope tube it will be seen that this takes place when the inclination is so great that the light no longer traverses the plate until the other side of it begins to be presented to the object.

Continuing, the two images gradually come together till the plates lie once more at right angles to the tube, having turned through 180°. It should be noted that there is nothing to prevent the plates from being •rotated any number of times, and that there are four possible positions of the plates to each displacement, with nothing but the readings of the divided circles to show which of the four positions you are dealing with.

A complete table has therefore to be made, giving the four positions for each millimetre of displacement.

Precautions.

Backlash. Read what is said on this subject under 'Calibration of the Spectroscope'. Approach all adjustments from one direction.

Vernier. To each plate there is a divided circle graduated in degrees, and to each circle there are two verniers 180 degrees apart. The object of this is to compensate for any error in the centring of the circle or in the way it fits the bearings — the mean of two readings at opposite ends of a diameter giving the true angle through which the circle has been turned. Book each reading separately and put the mean of the four at the end of the line.

The verniers enable the angle to be measured to the tenth part of a degree (Fig. 9).

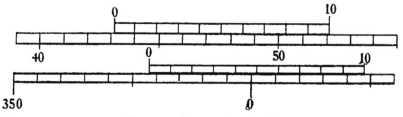

FIG. 9.   (43·1 and 355·7.)

On each of them is a short scale divided into ten parts, one end being marked for zero. But although the scale has ten divisions it is only the length of nine degrees, so that each division is nine-tenths of a degree long. Now suppose that the circle is set to exactly 32·1°, the zero of the vernier will lie between 32° and 33° without coinciding with either. But since 32·1 + 0·9 = 33·0 it follows that division 1 on the vernier will exactly coincide with 33·0°. But no other division on the vernier will coincide with any on the circle. Hence the reading is 32·1°. The same law holds for each fraction. Suppose the true reading is 44·7°. The zero of the vernier lies between 44° and 45°, and there is exact coincidence between No. 7 on the vernier and one of the degrees on the scale, but no other coincidence. For 0·7 + (7 × 0·9) = 7·0, so that as a matter of fact it is 44 + 7 = 51°, which coincides with o·7 on the vernier, although this is not of the smallest consequence, the only thing to note being which *vernier division* coincides.

Thus the *zero of the vernier* shows the *number of degrees,* and tne *vernier division which coincides* shows the number of *tenths of a degree.*

A warning which has not proved superfluous is that you should not take the wrong end of the vernier for zero. Zero is generally marked by a long stroke or an arrow-head. But the circle graduations and those of the vernier count always in the same direction, so that the zero is on the side of the lower numbers. Some-times there is an extra division on the other side of zero, making eleven in all. This is to facilitate reading if the zero comes just short of a degree, so that you need not shift to the other end of the vernier to see if it should read, say, 32·0 or 31·9.

If a magnifying glass is used, care must be taken to hold it at right angles to the graduated surfaces.

Plotting the Curve. The displacement being from o to 7 mms., and the inclination from o to something less than 80°, a very fair curve may be obtained by plotting 2 cms. to 1 mm. as abscissae against 2 mms. to 1 degree as ordinates. On the general subject of plot-ting, see Calibration, p. 107.

The curve should be quite uniform. Any observation that does not lie on it or would introduce a sharp bend, is wrong, and must be repeated. This power of detect-ing errors of observation is one of the great advantages of calibration curves.

The four positions are: from o° to 80°; from o°, i.e. 360°, backwards to 280°; from 180° to 260°, and from

180° backwards to 100°. Any one of these would suffice for measuring purposes, but as it is impossible to tell during an observation which one you are using, it saves time and temper to calibrate them all.

To measure the Radius of Curvature and Astigmatism of an Artificial Eye.

Two glow-lamps, about half a metre apart, are supported one on each side of the telescope by an arm clamped to its tube. The artificial eye is placed at a distance of about one metre from the telescope, which is then focused upon the reflected images of the two lights visible in it. The operation is best conducted in a dark or partly darkened room.

On turning the milled head the two images become four. If these are not exactly in a line they must be brought so by loosening the clamp and turning the arm carrying the lights. The milled head must then be further turned until the two nearest of the four images exactly coincide. When this is the case, read all four verniers and take the mean.

It is better to repeat this operation two or three times.

Then ascertain from your calibration curve what is the exact amount in millimetres and decimals of a millimetre corresponding to this scale reading. Call this B.

Then if R = radius of curvature,

y = distance between the two lamps,

x = distance of their plane from the centre of curvature,

$$R = \frac{2\,Bx}{y} \text{ approximately.}$$

This measures the radius of curvature in the line of
the two lamps. To detect astigmatism turn the telescope,
with the lamps, being careful not to touch the clamp
that holds them, about its own axis, and observe the
images. If they remain in contact there is no astigmatism. If they separate, or overlap more, proceed to
measure B in the directions of greatest and of least
separation. These will be at right angles to each other,
and there is a divided circle on the telescope tube by
which their position may be recorded.

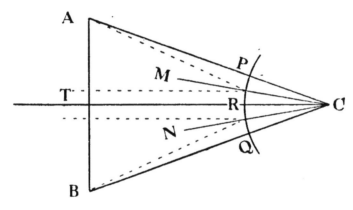

FIG. 10.

The difference between the reciprocals of the two
corresponding values of R gives the astigmatism in
diopters.

Theory. To prove that the displacement of the rays
is parallel and not angular, adjust the instrument to a
displacement of exactly 5 mms., with the scale one metre
off. Then remove it to the other end of the room. The
telescope will have to be focused again, but the displacement of the images will be exactly 5 mms. as before.

Since the angle of reflection equals the angle of incidence, if a reflecting surface turns through an angle a

the angle between the reflected and the incident rays is increased by $2a$. Moreover, any ray passing through the centre is normal to the surface when it cuts the sphere. Since for the light from A (Fig. 10) to enter the telescope the whole angle between the rays is ACT, the angle between the normals must be half this, namely, TCM, and the point at which CM cuts the sphere will be the point on which the reflected image will be seen.

In practice the angles ACT and BCT are equal, and since $TC = x$ and $AB = y$,

$$AB = 2 \tan (ACT) = \frac{AB}{TC} = \frac{y}{x}.$$

But the distance between the images is

$$B = R \times 2 \sin (MCT)$$
$$= R \times 2 \sin (\tfrac{1}{2} ACT).$$

Now when the angle is small, the sine and the arc and the tangent of it are sensibly equal. Assuming this, and putting the angle $ACT = 2x$, we have

$$\frac{y}{x} = \tan 2a, \ B = R \sin a,$$

so that approximately

$$B = R \times \frac{1}{2} \left(\frac{y}{x} \right), \ \therefore \ R = \frac{2 Bx}{y}.$$

If a more accurate result is desired, evaluate $\frac{y}{x} = \tan 2a$,

and find $2a$ in a table of tangents. Thence calculate a and find $\sin a$ in a table of sines.

Then $\dfrac{B}{\sin a} = R$, or as it is simpler to multiply, and

$\dfrac{1}{\sin a} = \operatorname{cosec} a$, find $\operatorname{cosec} a$ by the tables.

Then $B \times \operatorname{cosec} a = R$.

In place of an artificial eye a glass bulb will serve, if the interior reflections are got rid of by coating it inside with a dark varnish made by dissolving aniline purple in French polish. The curvature of such a bulb is seldom quite regular or quite spherical, but the diameter as calculated from observations on four sides will agree closely with the mean diameter as measured in the same two directions with the callipers.

II. 5. The Phakoscope.

The phakoscope is an apparatus for observing without the need of a dark-room the changes in the refracting surfaces of the eye during accommodation.

The person to be examined places his eye against an aperture in one side of a box which is black inside. In front of him is a hole through which he can see distant objects, and fixed across the hole a pin to serve as an alternative object. In the right side of the box are two small prisms arranged so as to reflect light from the window into the eye, and on the left side is an aperture through which the observer looks.

Three images should be visible. The first and brightest is given by the reflection of the two squares of light from the prisms in the surface of the cornea. The second, which is larger and not nearly so bright, is given by the front surface of the lens. The third, which is very faint, is from the back surface of the lens. It is smaller and is inverted, the other two being erect.

If now the focus of the eye is changed from a distant object to the nearest possible, the second pair of images will come nearer together and grow smaller, but the first and third will undergo no visible change, proving that during accommodation the front surface only of the lens is altered, becoming more strongly curved.

It is not always easy to see all three images in a phakoscope, unless the pupil is dilated by atropine, owing to the angle of incidence being too great. But the experiment may be easily made in the dark-room, using an incandescent lamp at a distance of a metre or less as object, and observing through the ophthalmo-meter. The brightness of the light makes it easier to see the very faint third image, and from its shape the observer can at once recognize which of the images are erect and which inverted. Moreover, with the ophthal-mometer it is possible to *measure* the change of curvature.

*II. 6. Acuity of Vision—Central.
(*Burch's Apparatus.*)

Two fine wires about 0·1 mm. in diameter are stretched across a pair of sliding frames furnished with a screw adjustment so that they can be set parallel to each other at any desired distance from contact up to 3 cms.

The frames are mounted in a ring which can be rotated behind a circular aperture in a screen so that the operator can alter the direction of the wires without affording any clue to the observer.

A sheet of white blotting-paper is fixed at an angle of 45° a foot or more behind the screen so as to receive the light of the sky and serve for a white background against which the wires appear as black lines.

1. Ascertain the maximum distance at which the

observer can distinguish the wires when they are $2\frac{1}{2}$ or 3 cms. apart. To make sure that he really does see them the operator alters the direction two or three times, and the observer indicates what he believes it to be with a short stick.

Call the distance l, and the diameter of the wire d. Then $\frac{d}{l}$ = the chord of the smallest angle of black upon white that produces a visible effect.

The chord of 1 second being $0.00000485 = \frac{1}{206264}$, it is easy to express $\frac{d}{l}$ in seconds.

This does not, however, measure the visual acuity. Owing to inevitable irregularities of refraction in addition to the errors of spherical and chromatic aberration, the image of a point is not a point but is spread over an appreciable area. And the wire is visible, not when its geometrical image is large enough to be discerned, but when the black of it, mixing with the white, and spread over that area, makes a perceptible shade of grey.

2. Having ascertained the maximum distance l at which a single wire is visible, the observer goes to a distance $\frac{3l}{2}$, half as great again. He is now quite unable to see the separate wires, but when the operator by turning the screw brings them within a certain distance of each other, they suddenly appear, not as two, but as one wire. Let the distance between the wires when they thus become visible = b.

Then $\dfrac{b}{\dfrac{3l}{2}} = \dfrac{2b}{3l} = V$ is a measure of the visual acuity.

Visual acuity is greatly diminished in a feeble illumination. It is also influenced by the colour of the light.

Place the apparatus in the dark-room. Illuminate the blotting-paper by a standard candle at successive distances of 1, 2, 3, 4 metres, and repeat the previous measurements.

Place a red glass in front of the candle and repeat.

Do the same with blue glass instead of red. This will give very low values for the acuity.

A better plan is to project a spectrum on to the blotting-paper. The influence of colour is then shown in a striking manner.

II. 7.   Sight-testing.

The power of perceiving signal lamps at great distances and distinguishing their colour is a form of visual acuity combined with colour-sense that is of considerable technical importance. Various methods of testing it are prescribed by the governments of different countries. The method described here has no official authority, but is of interest as reproducing in the laboratory the optical conditions under which the lights have to be recognized in practice.

The novelty of it consists in the use of an inverted telescope to produce the appearance of gradual approach from a distance, and a diagonal mirror on the principle of Pepper's ghost to imitate the disturbing effect of twilight or sunset.

The source of light is an ordinary oil-lamp—or electric arc, if lighthouses are to be represented. In front of this is a frame in which sundry plates, pierced with holes arranged to represent the lights of a ship, can be placed. Red or green glass may be interposed over any of the holes as required.

A reduced image of such a set of 'signal lights' is formed by a lens of rather short focus—say one inch, and received into the eyepiece of the telescope, which may conveniently magnify five times.

The telescope is mounted on a long slide with a range of at least five feet, at the end of which is a $\frac{1}{4}$-plate camera lens by which the final image is formed.

What happens is easily explained. Objects seen through a telescope from the large end of it appear magnified if quite close, but very greatly reduced when moved to a distance. That is because in any telescope focused for infinity, space is magnified longitudinally in the proportion of the square of the lateral magnifying power. So that if the telescope magnifies 5 times, distances are magnified 25 times, and with a power of 12 diameters distances are magnified 144 times.

A second characteristic of telescopes is that whereas with a projection lantern the farther the image is from the lens the more it is magnified, with a telescope the magnification is constant all along the axis, so that if the image is formed farther away *it looks smaller exactly* as if it were a real object. A third property common to telescopes and all systems of two lenses is that any object in the first principal focus of the first lens forms an image at the second principal focus of the second lens. This enables us to fix the point from which distances have to be measured.

Suppose the plate used contains two holes each of 1 mm. and 10 mms. apart. Let the telescope have an

eye-lens of 5 cms. and an objective of 25 cms. focal
length.   Let the first image, reduced ten times, be formed
at the first principal focus of the eye-lens.   It will form
a second image five times as large in the second principal focus of the objective, i. e. 25 cms. beyond the end
of the telescope.   This will be reduced about ten times
by the camera lens 150 cms. away, giving a final image
of say half a millimetre between the holes.

Now if the telescope is moved away from the light
one centimetre, the image recedes along the telescope
25 cms., but at the same time the telescope approaches
the observer 1 cm., so that the total effect is as though
the object were removed 24 cms.   Since we have a range
of $1\frac{1}{2}$ metres, the effect is as though an object consisting
of a pair of apertures half a millimetre apart were
removed to a distance of 36 metres.   It is clear that this
would render it practically invisible.

It should be noted that the final image produced by
the camera lens is practically at a fixed distance. - To
imitate the effect of twilight, sunset, &c., a diagonal
mirror of unsilvered glass is used to reflect a trans-
parency representing the sea with a distant coast-
line, &c.   By suitably arranging the light, sunset colours
can be introduced.   The effect is very striking.   The
signal lights appear in the offing, barely visible at first,
and gradually grow brighter and more distinct until the
colours can be made out.   The effect of a slight fog can
be imitated by means of very finely ground glass in
front of and not quite close to the plates.

The distance at which the person tested can recognize
the light is read off at once by the position of the
telescope on the slide.

II. 8.   To measure the Distance between the Centres of Rotation of the Two Eyes.
(*Ryland and Lang.*)

The distance between the pupils varies with the degree of convergence. It is also affected by any abnormal direction of the axis of either eye. But the distance between the centres of rotation of the eyeballs in their sockets remains constant unless the eye is displaced by some growth.

When an object is viewed alternately with the right eye and the left eye, its position as seen against a screen at a greater distance appears to change. If the amount of this change is measured, and the distance from object to screen is known, we can form an equation containing two unknown quantities, namely the distance from screen to eyes, and the distance between the centres of rotation of the eyes.

Therefore, by making a second equation with the eyes at the same distance from the screen but a different distance between object and screen we can determine the unknown quantities.

The apparatus consists of a baseboard carrying at one end a rest for the forehead and at the other a horizontal scale with sliding index. On the centre line of the baseboard, about midway between the eye and the scale, is a hole D (Fig. 11) in which is placed a pin, and there is a second hole at E, nearer the scale, to which the pin can be moved.

Place the pin at E, and close the left eye. Move the sliding index on the scale till it appears exactly behind

the point of the pin.   Repeat, using the left eye, with
the right eye closed.   Let the two positions of the index
be *a* and *b*.

Shift the pin to *D* and take two more readings, *c* and
d, of its apparent position.

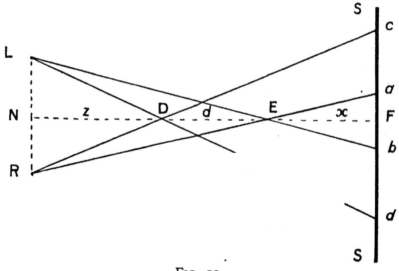

Fig. 11.

Let

$$DE = d$$
$$EF = x$$
$$ab = y$$
$$cd = y_1$$

These are the known quantities.
The unknown quantities are

$$ND = Z$$
$$LR = V$$

Then by a simple geometrical construction *V* may be
found.

Or, if *LR* be parallel to *SS*, we can calculate *V* as
follows :—

By similar triangles we have, with the pin at D

$\dfrac{V}{y} = \dfrac{z+d}{x}$, which may be written

$$\frac{Vx}{y} = z+d \tag{1}$$

and with the pin at E

$$\frac{V(d+x)}{y_1} = z \tag{2}$$

Subtracting (2) from (1)

$$V\left(\frac{x}{y} - \frac{d+x}{y_1}\right) = d$$

Whence $\qquad V = \dfrac{d(yy_1)}{xy_1 - y(d+x)}$

In the second form the apparatus (Fig. 12) has three pins k, l, m, in a row, parallel with the scale and about midway between it and the eyes.

The apparent positions of k and l on the scale at r and s, are observed with the left eye, and those of l and m on the scale at p and q, are observed with the right eye.

Putting $km = a$, $ps = b$, and $rq = c$ and putting $LR = V$ as before, we have

$$b : V :: (c-a) : (a-V)$$

whence $\qquad V = \dfrac{ab}{b+c-a}$

The value of V can also be found graphically.

The first method has this advantage, that the patient can be allowed to rest before taking the readings with the pin in the second position.

In both cases calculation can be avoided by con-
structing from the formula a table of values corresponding
to each pair of readings.

It is not necessary that there should be distinct vision

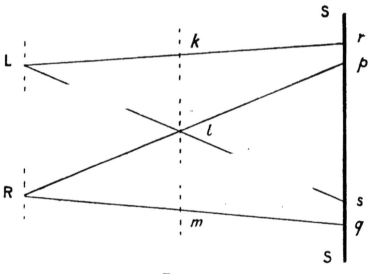

Fig. 12.

with both eyes at both distances, all that is required
being that the confusion circle of the point of the movable
index should be made concentric with that of the fixed
index.

SECTION III

JUDGEMENTS OF THE EYE: SPACE

Judgement of Distance.

The apparatus, consisting of two wooden trays one supported above the other, is arranged for two entirely distinct experiments.

*III. 1. Bourdon's Experiment.

In Bourdon's experiment three plumb-lines are suspended in a row at right angles to the line of sight. They are about 4 cms. apart, and the middle one works on a slide so that it can be placed nearer the eye than the others, or farther off, the amount of the displacement being read off on a millimetre scale.

The plummets hang in a basin of water or glycerine to prevent vibration, and screens hide the tops and bottoms of the threads so that there is nothing but the focusing power of the eye, if the observations are made with one only, or the stereoscopic sense, if both eyes are used, to enable the observer to judge their distance.

On the other hand, as the lines are perfectly still, any amount of time may be taken for careful observation. The position of the head should, however, be fixed.

An operator having set the middle plummet, the

E 2

observer states whether it is nearer or farther away than the outer ones, and roughly by how much.

A large number of experiments should be made, and the average correctness of the results recorded.

Accuracy will be found to depend largely on focusing power and the acuity of vision or defining power, combined.

*III. 2. Hering's Experiment.

In this experiment a momentary glimpse is given of a falling body, so that focusing and parallax are eliminated.

The same screens prevent the beginning and the end of the fall from being visible, and the lower tray is lined with wadding so that the impact cannot be heard.

A large pin, with a bead head, is placed at a suitable distance from the eye.

If the bead is white a black or dark background is used—if black, a white background.

There are a number of holes in the upper tray, at definite distances from the pin, some nearer, some farther off. Through these holes dried peas are dropped one at a time by the operator, and the observer has to say whether they are nearer or farther off than the pin. Things are so arranged that the peas never fall exactly in a line with the pin, as they might be seen in front of or behind it.

The operation is rather lengthy as only the average of a number of observations is of any use.

For example, take eleven sets, five nearer, five farther off than the pin, and one level with it.

Mark on a sheet of paper these positions and assign ten peas to each.

Let the operator drop one through, taking the holes at random. There are only two possibilities—either the observer is right or he is wrong. If his judgement is pure chance, the ratio of right to wrong will be unity, provided a sufficient number of cases is taken. If it is not unity, there must be some power of distinguishing distance.

Accordingly, having dropped an equal number of peas through each of the holes selected, and having recorded the judgements with a zero or a cross according as they were right or wrong, count up the results and give the ratio for each distance.

It is a good plan to rule a sheet of manuscript paper with a line down the middle to represent the pin and columns on each side of it, representing distances of 1, 2, 3, 4, 5 cms. nearer or farther off. Then, as each observation is made, the result is recorded in the proper column, taking care that the observer shall never know at what distance the next pea is to drop. Much better results will be got by taking twenty peas for each distance than with ten. The sheet of results with the ratios calculated, must be handed to the lecturer, signed with the name of the observer.

As two persons are required, each must act in turn as operator. But to avoid fatigue, and to ensure both getting a fair share of observation, they may change places between every fifty trials.

It is understood that as these experiments are intended to measure a personal faculty, care must be taken not to confuse the record sheets.

Observations should be made first with both eyes, because there is no need to repeat, when working with one eye, any of the experiments at distances which could not be distinguished with two.

III. 3. Donders' Isoscope.

This instrument consists of a square frame to which is attached, by the centres of its upright arms, a jointed parallelogram carrying a stretched thread at a distance of some 8 or 10 centimetres from another thread stretched vertically across the frame. The threads are first set parallel to each other by adjusting each to a plumb-line.

1. Bringing the face within 5 or 6 centimetres of the frame, look at an object 3 or 4 metres off. The two white threads should be visible as blurred parallel lines one on each side of the object looked at.

2. Without moving the head, fix the eyes on an object 30 or 40 centimetres off. The two white threads should now appear to coincide or nearly so, but they are probably no longer parallel.

Adjust the parallelogram till the threads are what you consider parallel. The actual angle between them may now be found as follows.

Let into the top and bottom rails of the frame are two short pieces of millimetre scale. They are exactly 50 cms. apart. One edge of the vertical side of the parallelogram crosses these two scales at zero when the threads are parallel. When the parallelogram is skewed we need only add the top and bottom scale-readings and divide by 50 to get the tangent of the angle of the skew—or what is simpler, double the sum of the scale-readings and divide by 100. The angle can then be found by a table of tangents.

The experiment should be repeated with the object (*a*) straight in front, (*b*) high up, (*c*) low down, (*d*) to the right, (*e*) to the left—keeping it always at the same distance from the eye, and not moving the head.

III. 4. The Wall-paper Experiment.
(*Tapetenbilder.*)

Use as object a wall-paper with some well-defined simple pattern, repeating itself every foot; or, better, every 6 inches. Placing yourself in front of it at some distance—3 or 4 yards—direct your eyes to some small object, e. g. a bead on a wire, that can be gradually brought nearer. The convergence of the axes of the eyes causes the more distant image of the paper to double, the separation increasing until the pattern over-laps by an entire unit, when quite suddenly the two images will appear to fuse into one much nearer than the wall and also much smaller than the original.

If you continue to move the object nearer, the images will again double, the overlap will increase until it amounts to two units of pattern, when there will be a sudden fusion and fitting together of the pattern which will look still smaller and nearer.

The process may be repeated until three or four over-laps have been noted. But by this time the observer will probably realize that he is squinting badly and that the effort is painfully fatiguing.

It is interesting to note how the mind, misled by the muscular effort of convergence, judges the distance of the images accordingly, taking no heed of the fact that the focusing does not correspond with the con-vergence. But it is this lack of correspondence that causes the sensation of fatigue.

The phenomenon is one that many children find out for themselves—the present writer remembers doing so when seven years old, and he has come across many other cases.

III. 5. Inclination of Zöllner's Lines.

A sheet of clear glass has painted on it two sets of bold parallel black straight lines, the one set sloping from left to right and the other from right to left, the two together constituting a series of capital V's one within the other from top to bottom of the glass, which they nearly cover from side to side.

If this sheet is placed over a pair of parallel lines so that all the black lines upon the one slope one way, and those upon the other in the opposite direction, those lines will not appear parallel, but inclined to each other.

Let Donders' Isoscope (see III. 3) be used to furnish the pair of straight lines, and adjust the movable one until in your judgement it appears parallel to the other. Then the actual angle between them may be measured in the usual way.

III. 6. The Müller-Lyer Paradox.

The central arrow-head and line are ruled on the fixed part of the apparatus, and the outside ones on strips of white card that slide over it.

One of the slides having been drawn out to the distance determined on, the other is drawn out until the point of it is at the same distance, to the best of your judgement.

The slides are graduated on the back so that the actual distances can be read off. Several attempts should be made, setting sometimes one and sometimes the other slide first, and varying the space between them.

Enter the results in tabular form, giving the percentage of error $\dfrac{b}{a}$ in a third column

a,	b,	$\dfrac{b}{a}$	$\dfrac{b}{a}$,	b,	a.

Arrange the horizontal lines in order of distance between the points of the fixed slide, taking 1, 2, 3, 4, &c., cms. For the first three columns let a be fixed and adjust b—for the last three columns let b be fixed and adjust a. The observations should not be made in the order in which they are entered. Two men working together should observe alternately, each recording the other's figures, but without looking at his observations and without letting him see the figures. This can be easily managed by putting the apparatus in the cover of a note-book and turning it over as it is passed from one to the other.

After reading the figures the second man should close the slides and pass it back to the first, who will then set one of the slides for the second man to make his observation.

Einthoven showed that the blurring of the lines, if a photograph of the figure is taken a little out of focus, produces an actual lengthening on one side and shortening on the other that corresponds in some degree to the illusion observed, and may possibly explain it.

III. 7. Estimation of a Right Angle.

According to Helmholtz (*Handbuch der physiologischen Optik*, 2nd edition, p. 687) most people, if they try to raise a perpendicular from a point on a horizontal line with one eye, make it lean over to the right hand when using the right eye, and to the left hand when they shut the right and open the left, to the amount of a degree or so. To try this experiment, rule a fine pencil line on a sheet of (unlined) paper. Make a pin-prick on any convenient part of it, and 8 or 10 centimetres away make another pin-prick at the point which you consider the perpendicular should pass through, keeping one eye shut. A pencil with a fine point may be used, but a pin or mounted needle is more accurate. Then rule a fine line through the two pin-pricks and measure the angle it makes with the other line by means of a horn protractor.

Repeat, using the other eye. Vary the conditions by dropping a perpendicular upon the line from a given point above it.

For some years I made all my students try this experiment, and from the results it was evident that the amount of the error depended mainly upon the education of the eye, but that the direction of it was entirely a matter of chance. Out of several hundred persons there were quite as many whose practice con- futed the theory of Helmholtz as there were who con- firmed it.

But the carpenters and fitters attending technical classes obtained remarkably accurate results—especially the older men.

III. 8. Eye Estimation of Fractions of an Interval.

It is an excellent thing to acquire the habit of esti- mating the decimal value of any given fraction of an interval, and with a little practice it can be done with some degree of accuracy.

The best plan is to divide the entire space in imagina- tion first into halves and then into quarters. From quarters to tenths is an easy step, because you can imagine the half divided into five parts, of which the middle one is bisected by the quarter division. At this point I generally visualize a scale divided into ten parts extending over the entire interval, running my eye backwards and forwards from end to end of it to get the imaginary divisions equal.

Suppose I want 37 per cent. I take the division 3 to 4 in my mind and drop straight on to the seven-tenths of it, fixing the point with my eye till I can mark it with a pencil. But the moment I fix the point the visualized scale vanishes. Having marked it I start afresh and visualize the scale again, this time trying to read off the decimal value of the point marked, correcting it if necessary.

Both the division of an interval in a given decimal ratio and the estimation of the decimal value of a given division should be practised. Many people habitually get within 2 per cent. But it makes them largely independent of some of the optical illusions concerned with estimation of length.

Examples: Find 0·07, 0·17, 0·25, 0·37, 0·53, 0·90 on lines of different lengths.

SECTION IV

SENSATIONS OF THE EYE

*IV. 1. Time-Relations of the Positive After-effect.

Apparatus required:—A simple photometer bench with
  candle or glow-lamp on a movable carriage, milk-
  glass screen, a box of red ' Bengal matches '.
  Three observers are required to get good results.

After dazzling the eye with a bright red light, if we
look at any white surface we see upon it a bright green
image of the flame. This is the well-known negative
after-image.

If the experiment is made in an absolutely dark room,
in which the observer has already remained for a while
that the effects of other light upon the eye may have
subsided, then the after-image of the red light is red.
This is the positive after-image discovered in 1786 by
Robert Waring Darwin, who showed that it is converted
into the negative image of complementary colour it
white light is admitted to the dark-room.

The experiment now described answers the question,
'How much white light is required to convert the
positive into a negative after-image?'

It should be attempted after some of the other experiments requiring a dark-room, e. g. retinoscopy, skia-

scopy, Prof. Gotch's spinthariscope, &c., so that the eye may be fairly well adapted for darkness.

1. Place the milk-glass screen at one end of the photometer bar and light the candle or incandescent light attached to the travelling carriage. See that the latter runs easily.

Sitting behind the milk-glass screen, the observer lights one of the Bengal matches and looks fixedly at it until it ceases to burn red, when it should be immediately dropped into a bowl of water so that no light of any other colour may affect the eye.

After about two seconds he glances at the milk-glass screen. If a green after-image is seen on it, he runs the light rapidly away so as to reduce the illumination until the green gives place to the red positive after-effect.

One colleague reads off the position of the light and the other notes the time to the nearest second.

The observer, who should shut his eyes immediately after making an observation, now opens them again and finds that the after-image which had appeared red has become green once more. He immediately moves the light still farther away until the red re-appears, the time and position of the light being again read by his colleagues.

Everything depends on the regularity and rapidity with which these operations are carried out. The adjustment can be made with some accuracy because owing to the flaming of the match the borders of the after-image are less strongly excited than the central portion. Consequently, at the moment of transition from negative to positive as the light on the screen is weakened, the central portion changes first and the green image appears suddenly to develop a black centre. This should be taken as the criterion, and the hand should stop and the eyes be closed the instant it is perceived. At first, readings may be taken almost as rapidly as they can be written down, but soon longer intervals of rest must be allowed.

If the eyes are not closed the after-image has a tendency to wax and wane in brightness in a troublesome manner.

With a very little practice twenty or more readings may be obtained in the three minutes or so during which the phenomenon is bright enough for this purpose.

2. The results should be studied by plotting the intensity of light as ordinate against time in seconds as abscissa.

The intensity of the light varies inversely as the square of the distance of the candle from the middle of the milk-glass screen. In addition to the readings taken during the experiment it is necessary to know the scale-reading when the unlighted candle is in contact with the screen, so that the actual distances from screen to candle may be calculated.

Most people fail more or less in the first attempt, but succeed as soon as they have mastered the routine. As the matches often strike badly a small spirit-lamp or Bunsen flame should be provided, screened from view except on leaning back 6 inches or so. The observer should avoid sudden movements, especially of the head, and he should look at the screen only just long enough to make an observation.

All the readings are made and recorded by the other

two, the observer merely saying 'Now' as soon as he completes an adjustment.

By taking turns to observe, the eye gets rested for another experiment.

While the match is burning the man whose turn is coming next should cover his eyes, and the third man, who has charge of the watch, must keep his eyes on that, and sit where the direct light will not affect him.

Brown-paper shades for the eyes may be worn by both with advantage.

It should be noted that the presence in the dark-room of light-coloured objects by which light from the photometer lamp may be reflected on to the milk-glass screen will seriously affect or even vitiate the results. A black velvet screen round the lamp will be of great service, and black gloves and coat will enable the observer who takes the scale readings to do so without delay as well as without risk.

*IV. 2. Measurement of the Negative After-effect.

These experiments may be made with a grease-spot photometer or, better still, a paraffin block photometer, but the best of all is a stereoscope with open ends, so that the lights to be compared fall each upon a strip of white card at an incidence of 45°. One card occupies the upper portion of the field of view of the right eye, and the other the lower half of the field of the left eye. A special 'fixation' star' enables the observer to keep the two images together without either gap or overlap.

With either of the other forms of photometer a black card must be fixed edgeways in such a position that the right eye sees only one side of the block or grease-spot and the left eye sees only the other.

Suppose the negative after-effect is to be measured in the right eye.

Set the left-hand lamp to some standard distance— say 50 cms.—and alter the distance of the right-hand lamp until the two balance.

A low-power microscope with large field of view having been arranged near a screened lamp so that its field of view is brightly and evenly illuminated, look steadily through it with the right eye at some small inconspicuous objects for a definite time—say from one to five minutes. Then look into the photometer and quickly increase the distance of the left-hand lamp until the two balance. Do this as quickly as possible, looking no longer than is necessary. Repeat every two minutes or thereabouts, the distances and times being recorded by a colleague.

Calling the distances d_1, d_2, &c., and the times T_1, T_2, &c.—reckoning from the moment the eye was taken from the fatiguing light—plot the values of $\frac{1}{d^2}$ vertically, against those of T horizontally. This will give the curve of recovery of the negative after-effect. It can be traced much farther than that of the positive after-effect. The way in which its shape is affected by the duration of the fatigue should be noted.

Some difficulty will be found in balancing the two luminosities, owing to the marked change in the apparent colour of the light after fatigue.

It is important to observe that each colour sensation is fatigued to a different extent. Hence the value of the following additional experiments.

Repeat the observations, using successively the red, green, blue, and violet screens, employing in each case the corresponding colour to produce the fatigue. Plot the results separately. It will be found that the fatigue is much greater and the recovery much slower after exposure to light of shorter wave-length.

Note the relation of these results to Purkinje's phenomenon.

N.B.—The influence of the size of the pupil may be eliminated by observing the photometer through a pair of pin-holes—one to each eye.

IV. 3. Experiments on Successive Contrast with a One-Prism Spectroscope.

A. *The Complementary Colours of the Spectrum.*

If the spectroscope is furnished with a reflected scale, take out the scale and throw light from the sky or from a white surface into the tube with a mirror so that it may be reflected from the second surface of the prism into the telescope, filling the field of view with diffused white light. Having covered the end of the tube with a black card, arrange the slit so as to show a continuous spectrum of considerable brightness. The entire spectrum should be visible at once, but the effect is more striking if a short slit is used so that it only covers about one-third of the field vertically. After looking steadily at this spectrum for a minute or so, keeping the eye fixed on the junction of the cross-wires, cover the slit with a card held in the left hand, and with the right take away the card covering the side tube.

A complete spectrum of complementary colours will

be seen across the white light that now fills the field of view.

If daylight has been used, the complementary to red will be blue, but if candle-light was used its complementary will be green. •

If the spectroscope is not furnished with a tube for a reflected scale, a cardboard tube, supported by a retort stand, will serve the purpose, all that is required being some means of reflecting light from the second surface of the prism into the field of view and of shutting it off until required.

B. *To show the effect upon the Spectrum of fatigue by Monochromatic Red Light.*

Substitute for the ' scale ' of the scale tube a cap with a horizontal slit ⅛th inch wide extending right across the tube. Illuminate this with the brightest light at your command, and cover it with red glass so as to have in the field of view of the spectroscope a narrow strip of bright red light extending across the field.

Arrange a continuous spectrum with fairly long slit so as if possible to cover the field. This spectrum must not be too brilliant. The Fraunhofer lines, if present, will serve for reference.

Cover the slit of the spectroscope and look fixedly for a minute or so on the centre of the band of bright red light.

Then with the right hand place a card over the slit of the side tube, and with the left hand uncover the slit of the spectroscope.

For a few moments an intensely black band will

appear across the red of the spectrum, ending in a green band in the part that was orange and yellow.

Everything depends, in this experiment, on the proper ratio of luminosity between the red light causing the fatigue and the red of the spectrum subsequently observed.

As a variant to this experiment, the spectroscope may be set up in the dark-room and the eye fatigued by looking at a Bengal match. The after-image of the red light seen against the red of the spectrum is black against the orange, it is dark green and bright green against the yellow.

But owing to the rapid changes in the intensity of this after-image, there is apt to be some confusion between the positive and the negative after-effects.

Moreover, there is a good deal of blue in the spectrum of the Bengal match. The effect of this may be prevented by holding a red glass between it and the eye.

C. *To show that light of any wave-length fatigues the eye strongly for light of the same wave-length.*

a. Set up the spectroscope in the dark-room and with a moderately wide slit arrange a lamp or two or three candles 8 or 10 inches away from the slit so as to give a continuous spectrum of merely moderate brightness. Then between the lamp and the slit place a Bunsen burner and observe some bright-line flame spectrum, fixing the eye steadily on one part of the spectrum. As soon as the platinum wire is removed from the flame the bright-line spectrum will be succeeded by a dark-line spectrum exactly corresponding to it, and so very definite and dark that it is hard to realize that this is merely an after-image and entirely subjective.

b. Focus across the slit of the spectroscope the filament of an electric lamp two or three feet away, with a lens, so as to show a narrqw band of great brightness in the spectrum. Place a candle near enough to the slit to give a spectrum of moderate intensity.

Look steadily at the bright band, and after a minute screen that light with a card. A uniform black band will appear to extend across the spectrum.

IV. 4. Alternation of After-images in the Stereoscope.

This experiment will not succeed unless the eye is to some extent dark-adapted. As it only occupies a short time it may be very well taken after Weber's **Law** (IV. 11) or the measurement of astigmatism with the ophthalmometer (II. 4) or any of the exercises involving a stay in the dark-room.

The apparatus consists of an ordinary stereoscope having as object a black card with a narrow diagonal slit in front of one eye, and a similar slit inclined the other way before the other. There is a short notch in the middle of each to act as fixation point.

The two slits combine binocularly to form the letter X.

Hold the stereoscope up to the sky and look steadily at the fixation point of the X for about three minutes.

Go into the dark-room and watch the after-images. First one will fade while the other grows brighter. Then the process will be reversed, and so on, the images being only visible together when at about half their maximum brightness.

One person should observe while another notes the times of equal brightness and of complete disappearance for the right eye and for the left. Sometimes the phenomena continue for a long while.

Repeat, first with a red glass over the slits and then with 'signal green'.

Finally, try red over one slit and green over the other.

IV. 5. Binocular Contrast.

Hering's Experiment.

An ordinary stereoscope is very convenient for this purpose, a square of red glass being inserted on one side of the central partition, and a square of blue glass on the other—or a plain box divided down the middle with a pair of eyeholes at one end and the coloured glasses at the other will do equally well, no lenses being required.

A small black wafer is fixed at the centre of each glass, with a white wafer close to the left side of the one on the right-hand glass, and another close to the right side of that on the left-hand glass. The two black wafers, being fixed binocularly, combine to form a single image, with the white wafers, each seen with a different eye against a different colour, one on each side of it.

If preferred, the white wafers may be one above and one below the black spot.

The instrument may be directed to a sheet of white paper, or to a mirror reflecting the light of the sky.

To bring out the full effect the light upon the white wafers must not be too bright. It should be reduced until the contrast colours show up strongly, orange against the blue glass and blue-green against the red.

The point brought out by Hering is this :—

Helmholtz had said that contrast effects were judgements of the mind. Hering shows that if you look with both eyes at once, though the colours of the two fields may combine, as they do with many people, to produce the impression of purple, the two contrast colours will not combine to produce the complementary yellow-green unless they too are superposed, but will retain each its own hue. It is difficult to imagine the mind combining two impressions, and at the same time keeping separate the judgements they give rise to. Hering argues very properly that contrast phenomena are functions of the retina rather than of the brain.

IV. 6. Spectroscopic Examination of Colour Contrast.

Burch's Experiment.

According to Hering the colours are associated in pairs, red with green and yellow with blue, so that after stimulation by one the other is naturally excited. Thus after excitation by red light, which uses up the red-green substance, there is a period of enhanced activity in the secretion of it, giving rise to the sensation of green. The validity of this hypothesis may be investigated by applying spectrum analysis to the colours produced by contrast.

Thorp's transparent replicas of Rowland's diffraction gratings afford an excellent means of doing this. They are practically transparent, and having 15,000 lines to the inch, give a spectrum of light coming from a slit out of the direct line of vision, with about as much dispersion as an ordinary good pocket spectroscope.

Over each eyepiece of an ordinary stereoscope is fixed a Thorp's replica of a diffraction grating. Two adjustable slits are held in a frame in front of the aperture by which light is admitted when viewing opaque photographs. The spectra of the first order of these two slits appear each near the middle of the corresponding glass, one a little to the left and the other to the right of it.

To prevent the colours of these spectra from being overpowered by the background, two black patches of exactly the right size are pasted on each of the coloured glasses. These patches are combined stereoscopically, and on one of them appears the spectrum seen by the right eye, and on the other that seen by the left.

The field of view appears to some persons purple, but to others the colour seems to oscillate between red and blue through purplish grey. But the two spectra look quite unlike each other. That seen against the red background shows little or no red but a splendid green and an equally splendid violet, while the other, seen against blue glass, has the red well developed, the green pale and dingy, and the blue almost absent.

It is important to have the slits of the right width, so that the colours of the spectra may be neither too bright nor too dull. The intensity of the effect increases as the observation continues.

A great deal depends on the kind of blue used. With cobalt glass the red is not very bright, owing to the transmission of a good deal of red by the cobalt. A gelatine film stained with prussian blue, in addition to the cobalt, greatly improves the red. A pale yellow film, which cuts off the violet, causes the violet of the spectrum to stand out strongly.

A magenta film brings out spectral green as the complementary colour. The physical stimulus is in this case complex, consisting of red and violet. On adding a yellow film, and thus cutting off the violet, the violet is added to the green of the contrast spectrum, but, if a blue glass is added instead of the yellow, the violet vanishes and the red stands out strongly with the green.

Thus it is shown that the complementary to any one simple colour is not *one* colour, but *all the rest* of the colours of the spectrum—a fact strongly opposed to Hering's theory, and in favour of that of Thomas Young.

IV. 7. Gotch's Ophthalmic Spinthariscope.

In Ramsay's spinthariscope a minute quantity of radium is brought near a card covered with powdered zinc blende. On examining with a lens the phosphorescent spot produced on the zinc blende it is seen to be caused by a constant succession of flashes, due, it is believed, to the impact of the α-rays of the radium.

To the physiologist these rays afford a retinal stimulus of minimal character and of practically constant intensity, which can be regulated in amount. Being minimal it does not produce retinal fatigue, and is imperceptible until the eye is dark-adapted to a certain extent.

Gotch's spinthariscope is furnished with the following adjustments :—

To regulate the density of the cloud of flashes, the phosphorescent screen can be brought nearer or moved farther from the radium by screwing or unscrewing the cap at the back of the apparatus. After any such adjustment the lens must be focused anew by screwing round the eye end of the tube. The screw slide at the side of the instrument serves to draw the radium to one side, thus reducing still farther the frequency of the flashes.

In order to limit the area over which the flashes are visible, a sliding diaphragm is provided. This must be kept closed whenever the instrument is not in use, as the daylight would make the screen phosphoresce with a steady glow, thus spoiling the effect of the radium. It has several apertures, some large and some small, the last being used to study the peripheral regions of the retina.

1. *General Phenomena.*

On first going into the dark-room nothing will be visible. After a few minutes, on looking straight into the spinthariscope, an ill-defined flashing luminosity will be just perceptible.

Direct the eye inwards towards the nasal side, the scintillations, viewed indirectly, appear brilliant and quite clear. Direct the eye outwards, the scintillations disappear. This shows that the nasal region of the retina is in a condition of much greater excitability than the temporal region.

By directing the eye upwards or downwards it may be shown that the sensitiveness to excitation of the corresponding retinal regions is in general greater than that of the temporal region, and greater than that of the fovea, but less than that of the nasal region.

In this respect there seems to be some variation in

different individuals, possibly depending partly on the conditions of exposure to light during the previous part of the day.

2. *Fatigue of the Observing Eye.*

Cover up the non-observing eye, and fatigue the observing eye by looking at daylight (the sky or white paper) for a few seconds.

Repeat the previous experiments.

With the fovea, the scintillations are imperceptible. With the nasal region of the retina they are perceptible as before. The effect of fatigue lasts longer in the fovea than in the periphery.

3. *Influence on the Observing Eye of Fatigue of the other Eye.*

Arrange the instrument so that the scintillations are barely visible to the dark-adapted eye.

Cover that eye, and illuminate the other with bright daylight for 10 or 15 seconds. Then in the dark repeat the observations. The scintillations will appear very brilliant to the eye that was covered, showing that the excitability of its fovea is increased by stimulation of the other eye. The effect on the peripheral regions is slight. The whole effect passes off in 10 to 15 seconds.

IV. 8. The Vessels of the Retina.

1. Looking straight before you at the black wall of the dark-room, let some one wave a lamp or candle up and down with a quick short stroke a foot or so

away on a level with the eye, but so far to the side that you can barely see it, if at all, without looking to that side.

After a few moments you will see dark shadows like the branches of a tree reaching from the periphery towards the centre of the field of view. These are the shadows of the vessels. If the movement of the light is up and down, the horizontal vessels will be seen; if to and fro, those that are vertically directed.

2. Stand as before, facing the black wall of the dark-room, and let some one with a lens of 2 inches focal length concentrate the light of a distant lamp on the sclerotic near the edge of the cornea. The lamp should be distant, because the smaller the spot of light and the brighter it is, the more perfect will be the view of the retinal vessels. A glow-lamp 10 or 15 feet off will serve.

Not only the larger vessels but even the capillaries can be seen in this way, and under some conditions the depression in the retina known as the yellow spot is rendered visible.

3. In the course of your spectroscopic experiments you may have occasion to note that when working on the violet end of the spectrum the circulation of blood in the eye often becomes subjectively visible, the separate corpuscles appearing as specks of light moving aimlessly about.

If, when these appear, you fix the eye on the cross-wires and watch one of these bright specks without looking at it, you will see that it pursues a definite path and is followed by others along the same path, which is in fact a capillary.

IV. 9. The Yellow Spot.

This experiment will not succeed with a person who has just come in out of bright sunshine. It requires the eye to be rested or, better, even somewhat dark-adapted.

A solution of chrome-alum in a glass tank or flat bottle is required. After covering the eye for a few moments with the hand, look suddenly at the sky through the solution of chrome-alum. If it has been made of the right strength you will see a round or oval magenta or crimson patch on the dull blue-violet of the rest of the sky. This patch is the area of the yellow spot.

Its angular magnitude may be measured as follows :—

Fix one of the rings of a retort-stand in a universal holder, so that it comes between you and the sky as you sit to make your observation. Set up another ring, on a stand clamped to the table, for you to look through. Adjust the distance of the first ring until it appears ot exactly the same size as the magenta spot.

Let the diameter of the first ring = d, and the distance between the two rings = r. Then $\dfrac{d}{2r}$ = the tangent of half the angle subtended by the ring. Find the angle by a table of tangents. If the spot appears oval, tilt the ring so as to make it appear elliptical, and measure its apparent diameter as seen from the fixed ring by getting some one to hold a centimetre rule behind it.

Chromium chloride may be used instead of chrome-alum.

IV. 10. The Blind Spot.

Place two small spots the size of a pea about 5 cms. apart on a sheet of paper. They may be black on a white ground, or white on a black ground, as is convenient. Observe them from a distance of 30 cms. or thereabouts. On closing the right eye and looking at the right-hand spot, the left-hand spot will vanish. The finger or the point of a pencil will remain visible while it is passed round the spot, but the spot itself will be completely invisible, if the eye is at exactly the right distance and in the right position.

By closing the left eye and looking at the left-hand spot, the right-hand spot may be made to vanish in the same way.

2. A more striking, but more difficult, way of performing the experiment is as follows :—

Place two lighted candles on a level with the eye, about 1 metre apart, in the dark-room, fixing black screens in front so that all but the two flames may be hidden.

Observe them from a distance of about 3 metres, holding edgeways between the eyes a large black card so that the left-hand candle is hidden from the right eye and the right-hand candle from the left eye.

If the distance and direction have been correctly chosen, when the eyes are directed to a spot midway between the two candles both flames will disappear.

3. The explanation is that under these conditions the image of the spot or flame falls upon that part of the retina where all the nerve fibres unite before passing through on the way to the brain.

Now, the nerve fibres are not themselves sensitive to light but merely transmit the excitation they receive from the retina. The spot, therefore, where they are all gathered together is blind.

The proof was given by Donders. Using an ophthal-moscope with a plane mirror, a small image of a light several metres away is formed by the eye itself, and this may be directed on to the optic nerve, which is seen to be lit up by it at the same moment that the light seems, to the subject, to have vanished.

*IV. 11. Weber's Law.

According to Weber's Law, the increase of stimulus required to be perceptible as an increase bears a con-stant ratio to the total stimulus.

For example, if you can just see a difference of 1 per cent. in a weak light it will require a difference of 1 per cent. in a strong light before you can distinguish it. You may be able to tell the light of 99 candles from that of 100, but you cannot tell 999 from 1,000—the least change you can detect would be given by 10 candles.

Fechner, expressing this law in terms of the differential calculus, by a very elementary application of the process of integration arrives at the law that—

' The intensity of the sensation varies as the logarithm of the stimulus '.

- By the following experiments it may be ascertained how far in your own experience these statements hold good.

At the end of a long graduated photometer bar is fixed a lamp surrounded by a black screen on the side next the observer, who sits so that he can look over it.

A number of blackened stands are provided, each carrying on a blackened wire a disk of white cardboard. These stands are arranged on the table at different distances, so that to the observer they appear to form a group above the bar fairly close together, care being taken that the nearer ones cast no shadow on those farther off. The distances are to be adjusted so that the illumination may appear to increase by equal stages from one disk to the next.

Thus, if A is farthest off, $A : B = B : C = C : D$, and so on.

One of the disks may be fixed to the sliding carrier of the photometer, and one or two of the nearest may be moved by hand, so that as many as five can be managed without much difficulty by an observer working alone. It is, however, better to have some one to help, in order that a larger number of disks may be set up. The disks vary in actual diameter, but should be arranged so as to appear of the same size, the larger ones being placed farther off. A good plan is to set up one at say 2 metres, and another at 1 metre, and then insert one, two, or three so as to divide the increase of illumination into what you consider two, three, or four equal stages. Most people find it easier to divide the interval into the larger number of stages.

Calculation.

On any surface normal to the light, the intensity I of the illumination varies inversely as the square of the distance d and directly as the candle-power K of the lamp—

$$I = \frac{K}{d^2}.$$

Having arranged the disks to suit the judgement of the observer, the distance of each from the light must be measured. This may be done for the more distant cards by means of the graduations of the photometer bar, using a set-square to determine the positions of the cards. For the nearer ones care must be taken to get the actual distance in a straight line, and unless both lamp and disk are symmetrically situated with regard to the bar, the scale readings will be incorrect.

The presence in the dark-room of persons in light clothing will vitiate the results—so will tobacco smoke, the inverse square law being only true of a clear atmosphere.

Let a = distance of A from the lamp

$b =$ „   „ B „   „   „   &c.

Then $\frac{1}{a^2}$ is proportional to the illumination on A,

$\frac{1}{b^2}$ „ B, &c.

By Weber's Law $\dfrac{\frac{1}{b^2}}{\frac{1}{a^2}} = \dfrac{\frac{1}{c^2}}{\frac{1}{b^2}}$, i.e. $\dfrac{a^2}{b^2} = \dfrac{b^2}{c^2}$, &c.

So that $\dfrac{a^2}{b^2} = \dfrac{b^2}{c^2} = \dfrac{c^2}{d^2} = \dfrac{d^2}{e^2}$, &c.

This is much more easily calculated by using logarithms.

Find $\log a$; double it $= \log a^2 = 2 \log a$.

Then $(2 \log a) - (2 \log b) = \log \dfrac{a^2}{b^2}$,

$$(2 \log b) - (2 \log c) = \log \frac{b^2}{c^2},$$

$$(2 \log c) - (2 \log d) = \log \frac{c^2}{d^2}.$$

If Weber's Law holds, these ratios should be the same and therefore their logarithms should be the same.

Having found what in your judgement is the correct position for the cards, you may calculate what it ought to be thus :—

Suppose five cards are to be arranged, the nearest at 1 metre, the farthest at 2 metres. The illumination at 1 metre will be four times that at 2, and as there are five cards, there must be four equal ratios.

$$\log 4 = 0.6020$$
$$\log 1 = 0.0000$$
$$\text{Difference} = 0.6020.$$

Divide this into four equal parts = 0.1505.

$\log a^2 = 0.6020$	$\log a = 0.3010$	$\therefore a = 2.000$
$\log b^2 = 0.4515$	$\log b = 0.2257$	$b = 1.682$
$\log c^2 = 0.3010$	$\log c = 0.1505$	$c = 1.414$
$\log d^2 = 0.1505$	$\log d = 0.0752$	$d = 1.189$
$\log e^2 = 0.0000$	$\log e = 0.0000$	$e = 1.000$

The last column gives the distances in metres. It is obvious that there is no need, in using logarithms, even for this calculation. We might just as well have taken the logarithms of the distances at once and divided their difference into the required number of parts.

Having found the 'correct' positions by calculation, it is interesting to alter the arrangement so as to put one of the largest disks nearest to the lamp. This generally makes the gradations appear incorrect.

A **red** glass over the lamp, so that the light is practically monochromatic, gives more concordant results because it eliminates Purkinje's phenomenon.

Effect of Light Adaptation.

Fix a funnel of white card, blackened inside but white outside, over one of the shutters in the wall. Bring the photometer bar in line with it so that you can see the cards while sitting in broad daylight. Let your eye be an inch or two away from the small end of the funnel so that it may not cease to be affected by the light of the room.

Compare your results for a series of five disks with those obtained while you were in the dark-room.

In most cases the statements made by Helmholtz on this subject are confirmed. With any series of seven or more cards, the ratios increase towards both extremities.

IV. 12. Purkinje's Phenomenon.

Taking from among Holmgren's Wools the brightest scarlet skein you can find, select other skeins of green, blue, and violet which appear to you its equal in brilliancy, and in addition some which appear darker. Let this be done in the full daylight.

Then examine them by diminished light in the dark-room. The red will be found to lose more in luminosity than the blue, so that the green, blue, and violet will **now** appear the brightest.

It is essential that the composition of the light in the dark-room should be the same as that outside.

To ensure this, reflect light from the sky by a mirror placed on the window-sill through a diaphragm—either Aubert's or iris—in the shutter of the dark-room.

For moderate illumination the skeins may be held in the beam of light at some distance from the hole, remembering that the illumination will be evenly distributed only over a limited area around the axis of the beam.

The law that the intensity of the illumination varies inversely as the square of the distance from the source holds if from the point considered nothing is visible through the diaphragm save the light of the sky in the mirror, the 'distance' being reckoned from the point to the diaphragm.

But if the diaphragm is so large that the whole mirror is visible through it at the greatest distance and some of the surroundings at shorter ones, then the 'distance' is to be reckoned from the point to the mirror on the window-sill as regards the light from the sky. And any appreciably bright object visible beside the mirror is to be treated as a separate 'source' and its contribution added, due regard being had to its 'distance'.

From this it will be evident that the formula for the variation with distance of the illumination might be very complex save when the diaphragm is so small and the area utilized so limited that nothing is visible from any point within its boundaries save the light of the sky reflected from the mirror. *It is time thrown away when observations are made under any other conditions.*

For feebler illumination a small disk of white card a centimetre or two in diameter is placed in the path of the beam, and is used as a secondary 'source' of light, distances being measured from it to the object examined.

After being in the dark-room for some time the sensitiveness of the retina for blue and violet will greatly increase, so that if care is taken not to let the light from the white card disk reach the eye, blue and violet skeins, so dark as to be almost black by daylight, will appear brighter than the most brilliant red. Colours like 'geranium red' will look violet, the red component of them being invisible.

This experiment explains one which was formerly quoted in support of Hering's theory of the functional connexion between red and green. After being for some time in a photographic dark-room lighted by gas behind the usual red cloth and glass, if the gas is suddenly turned down so as to reduce the light considerably, it appears for a few moments no longer red but blue-green.

Under these circumstances the small percentage of blue and green light transmitted by the coloured medium produces a much more powerful effect than the red when the light is turned low. Examination in a strong light with a hand spectroscope will show how much green and blue, and even violet, is transmitted by many of the so-called ruby cloths and glasses.

SECTION V

MEASUREMENT OF COLOUR SENSATIONS

V. 1. Calibration of the Spectroscope.

INASMUCH as no two spectroscopes are alike, the readings convey no intelligible meaning unless expressed in wave-lengths. Even if the instrument has been graduated by the maker in wave-lengths the time spent in calibrating it is not lost, for on the one hand there is always the possibility that the adjustments may have altered, and on the other hand the observer becomes in the process familiar with the spectrum from end to end, and is able to make the measurements of his colour-sensations with greater ease and accuracy.

The Calibration Curve.

If sunlight is available, make scale-readings of the positions of the principal Fraunhofer lines *A, B, C, D, E, b, F, G, h, H,* and *K.*

If sunlight is not available, or if you do not know the Fraunhofer lines, proceed as follows :—

Sodium.

Put a small piece or two of sodium thiosulphate or carbonate or iodide into the open end of a short piece of hard glass tubing the size of a pencil, and support it horizontally in a retort stand so that it may just touch the edge of the flame of a Bunsen burner. In a minute or two when the tube gets hot the flame will be coloured orange, and the bright lines of sodium will be visible in a spectroscope directed towards it.

a. Focus the eyepiece of the spectroscope so that the cross-wires are sharply visible.

b. Focus the telescope of the spectroscope so that the bright sodium lines are sharply visible.

c. Move the eye slightly from side to side while looking at one of the bright lines. If it appears to move from side to side past the cross-wires, its image is too near, but if the wires seem to move past the line its image is too far off. When correctly focused the bright line and the cross-wires will keep together when you move your head. Until so focused it is of no use taking a scale-reading.

d. Back-lash must be guarded against. By back-lash is meant a certain looseness of fitting on account of which in making the adjustments the scale moves slightly before the image moves, or vice versa, with the result that the readings are different according as the last movement was to the right or to the left. Consequently, to avoid back-lash all the readings must be taken in one direction, e. g. let the line be to the left of the cross-wires and gradually bring it up to them. If it goes the least bit beyond, turn right back and bring it up to them again. It is only in this way that reliable scale-readings can be obtained.

e. Variations in the width of the slit must also be guarded against.

Place one of the sodium lines on the cross-wires and open the slit wide. If the slit has one jaw fixed and the other movable the line will increase in width *on one side only*.   In that case let all measurements be made to *the fixed side of the line*.   If the slit has both jaws movable the line will broaden in *both directions at once*.   In that case measure always to *the centre of the line*.

If the spectroscope will separate the sodium lines the scale-reading of each must be taken separately.

Lithium Chloride.

The red line of lithium may be taken next. For this purpose a drop of a strong solution of lithium chloride is taken on a platinum wire and held in the Bunsen flame.

There is a right way of doing this, and a wrong way, leading to endless trouble.

a. Each wire must be used for *one substance only*. To facilitate this the wires are fixed, not in glass rods, but in tubes, and a slip of paper on which is written the name of the element is inserted by the other end and sealed in.

b. The end of the wire is made into a little curl by winding it three times round an ordinary pin. Before doing so it should be made red-hot to soften it, untwisted if already kinked or twisted, softened again, and made straight by drawing it over a pencil under the thumb. Only, as the glass will not stand the strain of the pull, the wire itself must be grasped between the forefinger and the thumb of the left hand. Very bad kinks may be smoothed out of the platinum wire without breaking it, if it is annealed often enough during the process.

c. A drop of the solution having been picked up in the

curl on the end of the wire, it is brought into the flame in the following manner.   The wire is held so as to slope *point downwards* at an angle of about 30 degrees to the horizon, and is brought into the flame on the side next the spectroscope so that the *flame touches the middle of the wire first.* It is then drawn up so that the curl comes into the hot part.

If the curl is put into the flame first you will probably see the drop come out of it and run up the wire to the glass handle without ever being volatilized. So much is this the case that with some substances you may fail altogether to get the spectrum for this reason, especially if, as beginners generally do, you rest your arm on the table and let the wire point upwards.

By putting the middle of the wire into the flame first the solution is driven into the curl and forced to evaporate there.

It is almost impossible to do this single-handed ; two people should work together, taking turns to observe.

Thallium Chloride.

This is a bright line, but lasts a few seconds only. It is generally accompanied by sodium as an impurity. If the spectroscope is one of very high power it may be a little difficult to find. A good plan is to illuminate with a continuous spectrum, and bring a rather yellowish green into the middle of the field. As soon as the splutter of the evaporation is over, take a fresh dip. After a while the wire will get loaded and the spectrum will persist longer.

A method which answers both for thallium and lithium is to melt together boric acid and a salt of the metàl. The powdered product keeps well, and gives on a loop of platinum wire a bead which shows the spectrum of the metal for a long time.[1]

Strontium Chloride.

The red part of the flame spectrum of strontium is not used for calibration because it consists of bands rather than lines and is not well defined. But there is a very sharp line in the blue, which is of great value, occurring in a part of the spectrum where there are few lines. It is about midway between the place where green changes to blue, and the place where blue becomes violet. When it has once been seen there will be no difficulty in finding it again.

Potassium Acid Tartrate.

Potassium gives us the extreme red of the visible spectrum, and no one who can see the line is red-blind.

To those with normal sight there are two difficulties— one that the line is not very bright, and the other that it is liable to be overpowered by the presence of sodium. That is why the potassium hydrogen tartrate—or cream of tartar—is selected. If on examination it should be found to show the sodium line, it should be recrystallized from hot water, and the crystals dried between folds of filter paper. This, repeated two or three times, will get rid of practically all the soda.

If, then, the observer remains in the dark-room for some time before making the attempt there should be

---

[1] I am indebted to Mr. D. H. Nagel, M.A., for this recipe.

no difficulty in seeing the red line, which should be sought *beyond* what appear to be *the limits of the red by daylight.*

The next step is to make a trial calibration curve, and to do so we must determine the scale on which it is to be plotted.

Let us suppose that the extreme red, represented by the potassium line, is 486 on the scale, and that the farthest limit to which violet can be traced is 1300 of the scale—say 450 to 1350, i. e. 900 scale divisions in all.

Since the potassium line is 7697 and the extreme violet about 3900, we may take 7700 to 3900, i. e. a range of 3800 Ångström units for the wave-lengths.

Now the object is to get these two as nearly equal as possible, remembering that in practice you can only use simple ratios such as 1 : 1, 1 : 2, 1 : 5, or occasionally 1 : 4, but that is troublesome.

If we take 5 scale divisions = 1 mm. the scale will go into 18 cms., and 20 Å. U. to 1 mm. will bring the wave-lengths into a range of 19 cms.

If you had only squared paper 22 cms. x 16 cms. it would be better to take 25 Å. U. = 1 mm., so that the wave-length range would occupy 15·2 cms., which would plot very well against the 18 cms. of the scale-reading. But double this size is better for the final curve.

Having settled on the size, proceed to mark off a scale of wave-lengths horizontally, beginning with 7700 on the left hand and ending with 3900 on the right. Then mark off a scale of spectroscope readings vertically, beginning with 450 at the bottom and ending with 1350 at the top.

(In the case of the paper 22 cms. × 16 cms. above referred to, the wave-lengths would be marked off vertically and the scale-readings horizontally.)

Then for potassium, follow the line corresponding to a scale-reading of 486 to the point where it intersects the line corresponding to a wave-length of 7697, which is that of potassium, and mark the intersection by a cross.

Suppose that the second of the two sodium lines is 606 on the scale of the spectroscope, follow the line corresponding to 606 to its intersection with the line corresponding to 5891 on the wave-length scale and mark the intersection with a cross. And so on with all the lines measured.

Draw lightly with pencil a free-hand curve through the points thus determined.

You are now in a position to identify the Fraunhofer lines. Begin with an easy one, C. Follow the line 6564 on the scale of wave-lengths to where it is cut by your free-hand curve, and from that point, turning at right angles, follow the line of scale-readings to the edge of the paper and read off the number corresponding to it. Set the spectroscope to that scale-reading, and on looking through it you will see a strong black line somewhere near the cross-wires—at any rate it will be the most noticeable line in the neighbourhood. Bring it on to the cross-wires, take the reading, and plot it as another point in your calibration, correcting the curve so as to pass through it.

Proceed to find B in the same way.

The group b in the green is easy to recognize, but E is hard to identify and should not be looked for until after F and b have been added and the curve corrected by them.

H and K are seen as broad black bars in the extreme violet by direct sunlight. They appear in the arc as bright lines when a piece of chalk is drawn across the carbon, but G, although its whereabouts is seen in a moment, is difficult of exact identification.

The spark spectrum of hydrogen in a Plücker tube will, if a small induction-coil is available, give the red line C, the blue-green F, and an indigo line close to G, besides a fainter one farther in the violet.

The spectrum of mercury vapour given by an Aron lamp may also be utilized.

When all these lines have been identified and plotted on a fairly large scale—38 cms. for the range of wave-lengths will serve—the corrected curve should be drawn so as to pass through them all. Any break or *sudden* change of direction in this curve is proof of an error either in identifying or in measuring the line, or, as is not seldom the case, a mistake in plotting. When all such mistakes have been rectified the line should be inked in, using a ruling pen and a flexible ruler with lead back which can be bent into any shape and will stay bent.

To use the Calibration Curve.

1. To set the spectroscope to a given wave-length.

Follow the line of the given wave-length from the edge of the squared paper to the curve. Thence proceed at right angles along the line of scale-readings to the other edge of the paper and set the pointer, or cross-wires of the spectroscope, to the scale-reading there found.

2. To find the wave-length of any part of the spectrum indicated by the pointer or the cross-wires.

Take the scale-reading.

Follow on the squared paper the line corresponding to that scale-reading, as far as the curve. Thence

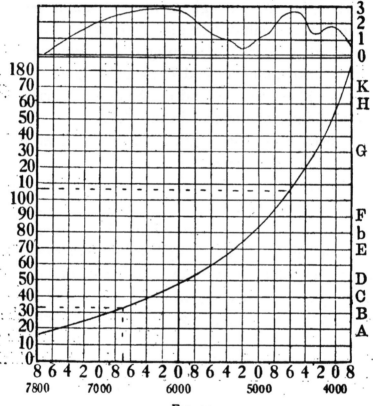

FIG. 13.

proceed at right angles along the line of wave-lengths to the edge of the squared paper and read the wave-length indicated there.

N.B.—If you are using a prism spectroscope of the ordinary type the calibration curve will be strongly curved at the ends. The red will be much compressed and the violet greatly drawn out.

The opposite is the case with a grating spectroscope.

The red is enormously drawn out and the violet somewhat compressed.

The calibration curve is practically a straight line, curved slightly if the grating is mounted on a prism.

It is well to work with both kinds, for practice.

NOTE.—The method of focusing the collimator has been omitted, because where the same spectroscope is in constant use by a number of people any alteration of the collimator would upset all their work. It should be focused once for all by a responsible person, and no one else should touch it.

The provision for effecting it differs in the various types of instrument. In general the telescope is first focused for infinity—on a star or a cloud, &c.

Then, keeping the telescope unaltered, the collimator is focused until it gives sharp images of the spectral lines.

The setting should be verified from time to time, otherwise it should not be disturbed.

V. 2. Wave-lengths of Lines used for Calibration.

The wave-lengths are given to the nearest Ångström Unit or 'tenth metre', the meaning of which is shown below.

The green line of thallium is   5351 Å.U.

which is equivalent to 0·0000005351 metre

i. e.   0·0005351 millimetre.

Hence, to turn Ångström Units, or 'tenth metres', into metres, put on the left as many ciphers as will make the number of decimal places up to ten. Hence the name 'tenth metre'.

Also 1 Å. U. = one ten-millionth part of a milli-metre. Thus the wave-length of this particular line in the green is rather more than $\frac{1}{2000}$ millimetre, or $\frac{1}{50000}$ inch.

Fraunhofer Lines.

Å. U.

$A = 7606$ In the extreme red
$B = 6869$ In the deep red
Hydrogen $C = 6564$ In the bright red
Sodium $D_1 = 5897$ In the orange
Sodium $D_2 = 5891$ In the orange
$E = 5271$ In the green
Magnesium $b_1 = 5184$ In the green
Magnesium $b_2 = 5174$ In the green
Magnesium b_3 and $b_4 = 5169$ In the green
Hydrogen $F = 4862$ Between green and blue
$G = 4308$ Between blue and violet
Calcium $H = 3969$ In the extreme violet
Calcium $K = 3934$ In the extreme violet

Hydrogen Vacuum Tube.

Å. U.

Bright   6562   In the bright red
Bright   4861   Between blue and green
Bright   4340   Near the indigo
Fainter 4101   In the violet

Bunsen Flame Spectra.

Potassium $\begin{cases} 7697 \\ 7663 \end{cases}$ Broad and difficult to see in the extreme red
Lithium   6707   Brilliant red

Sodium   5897⎫
Sodium   5891⎭ Orange, a double line
Thallium   5351 A yellowish green
Strontium 4609 Blue

Mercury.

(A globule of mercury enclosed in a vacuum tube.)

⎧5788⎫ Citron lines
⎩5768⎭ A pair
  5460 Green.   Very brilliant
  4357 Deep blue

*V. 3. The Spectroscopic Method of Measuring Colour Sensations.

This method, which has been described in my papers in the Philosophical Transactions, B. vol. 191, 1899, p. 1, and B. vol. 199, 1907, p. 231, depends on the fact that coloured light of quite moderate intensity rapidly induces a very noticeable degree of temporary colour-blindness if it occupies the entire visual field. This may be experienced by holding the eye near a large piece of scarlet cloth lying in the sun.

I therefore employ a spectroscope of considerable power, so that not more than one-tenth part of the spectrum is visible at one time. The eyepiece must be one having a wide field of view, and the spectrum, instead of occupying a narrow ribbon in the middle of the field, must fill it from top to bottom as well as from side to side. A good bold pointer in the centre of the field is a better reference mark than cross-wires, which inexperienced persons find difficult to see in the fainter portions of the spectrum at its extremities.

1. *Normal Colour-boundaries.*

The first step is to ascertain what names the person whose colour sensations are being measured employs to describe the colour of the spectrum in various parts. It does not in the least matter whether these are right or wrong, all that the operator wants to know being at which points, according to the observer, rapid changes of colour occur.

To a person of normal vision the spectrum appears to begin with a deep red of very feeble luminosity about λ 7600, gradually increasing in brightness without change of hue to a full red, which remains constant for some distance. At a certain point it becomes scarlet, changing rapidly to orange and then passing with great rapidity through yellow to yellow-green and more gradually to green, which continues unchanged over another considerable portion of the spectrum.

Quite suddenly this shows a trace of blue, and then passes through blue-green to blue with even greater rapidity than the change from red to green. Blue remains constant in hue for some distance, and then passes through indigo to violet, which ends about λ 3900.

2. *Fatigue Experiments.*

The second part of the test determines whether a change seen by the observer is really a change of hue to him, or merely a change of brightness.

This is done by measuring the change produced in the position of the junction of two colour sensations by fatiguing one or other of them.

To begin with, spectral yellow is produced by the admixture of equal quantities of red and green. If there is excess of red it becomes orange—if of green, we consider it greenish-yellow.

The observer is asked to select what is, in his opinion, the 'best yellow', i. e. neither orange nor greenish. He is then made to look for 30 seconds at a part of the spectrum where he can see no colour but red. The region between B and C will serve, but beyond C the proportion of green is too great.

The operator, holding the watch, begins at the 25th second to count 25, 26, 27, 28, 29, 30, and at the word the observer must traverse the spectrum rapidly till he reaches the part that looks the colour he had pitched on as 'best yellow'. Bringing this up to the pointer he must stop and take his eye away.

No second thoughts and no attempt at careful adjust-ment must be allowed, but the first momentary impres-sion must be recorded. After the scale-reading has been taken the observer may be allowed to look again, or to look with the other eye. He will probably, after seeing how rapidly the colour changes, enter much better into the spirit of the experiment, and record his momentary impressions.

If he possesses a green sensation as well as red the effect of fatiguing with red will be to make the 'best yellow' come farther towards the red, but if his 'red' was merely the beginning of green his 'best yellow', after fatigue with red, will move towards the blue, because in that case what he calls yellow is not a different colour, but merely a brighter green. The object of the fatigue experiments is to apply these tests systematically to the three chief junctions of colour sensation, namely, of red with green, of green with blue, and of blue with violet. It is essential that the same degree of fatigue should be produced in each case, and to ensure this a certain routine is followed.

For the green-blue junction, green up to, but not beyond, b should be used.

For the blue-violet junction, blue from F to about one-third of the way to G. After this, the observer should be asked to trace the violet to its extreme limits. This is easier done if the slit is shortened so that the spectrum only occupies a narrow strip across the middle of the field. During this process the blue will recover while the violet is fatigued.

The slit is then lengthened again and the eye fatigued with violet from $\lambda\,3900$ to about $\lambda\,4200$, and after 30 seconds the observer may find where blue begins.

Next fatigue with blue from G and about one-third of the way from G to F for 30 seconds, and then find where green begins. Observe that we cannot use the same blue when testing for green as we did when testing for violet, because then the presence of a little green with the blue did no harm, and now that we are looking for green a little violet with the blue will not hurt.

After this, we fatigue with green between F and b, but not on the red side of b, and look for the place where red begins. It may be noted that after working for a little while with the spectrum the idea of a blend between red and green seems to leave us, and we see—as Wollaston did—the two colours with the merest trace of yellow between them. This is due to an altera-tion, brought about by fatigue, in the shape of the two curves of colour sensation.

The last thing is for the observer to trace red as far

as he can see it, shortening the slit for the purpose as was done for the violet.

In each case the operator should call the times and record the scale-readings, verifying the colour by a hasty glance before doing so, as sometimes, especially in working back, an observer will run right past the colour he was looking for. Should this occur, cancel that observation and fatigue again.

3. *Interpretation of Results.*

The scale-readings so obtained having been, by means of the calibration curve, converted into wave-lengths, they are dealt with as follows, the example taken being that of an artist :—

First Stage.   Normal Colour-boundaries.

'Best Yellow', i. e. green meets red $= \lambda 5785$

blue meets green $= \lambda 4907$

'Indigo',   i. e. blue meets violet $= \lambda 4650$

Second Stage.   Fatigue Experiments.

After looking at—

Red for 30 seconds,	green begins at	$\lambda 6010$
Green ,, ,,	blue ,, ,,	$\lambda 5080$
Blue ,, ,,	violet ,, ,,	$\lambda 4740$
	Violet ends at	$\lambda 3900$

After looking at—

Violet for 30 seconds,	blue begins at	$\lambda 4555$
Blue ,, ,,	green ,, ,,	$\lambda 4720$
Green ,,	red ,, ,,	$\lambda 5550$
	Red ends	$\lambda 7600$

From these data we find—

1. The extent of the colour sensations.

Red extends from 7600, where red ends, to 5550, where red appears after green.

Green extends from 6010, where green appears after red, to 4720, where green appears after blue.

Blue extends from 5080, where blue appears after green, to 4555, where blue appears after violet.

Violet extends from 4740, where violet appears after blue, to 3900, where violet ends.

2. The extent of the compound colours :—

Yellow, where green overlaps red, from 6010 to 5550.

Blue-green, where blue overlaps green, from 5080 to 4720.

Indigo, where violet overlaps blue, from 4740 to 4555.

Observe that in this case violet and green overlap from 4740 to 4720.   This is unusual.

3. The middles of the overlaps correspond very closely with the Normal Colour-boundaries of the same person.

Thus, Red-green = 5780 instead of 5785
Green-blue = 4900 instead of 4907
Blue-violet = 4648 instead of 4650

Red-blind people have the spectrum shortened at the red end, and no *true* change of hue near the yellow. But they have the change from green to blue, and generally that from blue to violet, unless they are violet-blind also.

In green-blind people the spectrum is not shortened. They find no change of colour in the yellow, nor in the blue-green, but a very sudden and striking change in the middle of what we call green, at about 5180, where to us there is no change. As soon as this has been diagnosed, its existence as a change of hue should be verified by fatigue with light from D and from F in succession. This will affect it much more powerfully than the light used in testing for the red-green and the green-blue junctions. It is of course the red-blue junction, which is to most people completely masked by the green.

One of the most characteristic signs of green-blindness is that blue-green, between b and F, is very often called 'pink', or 'green', or 'grey', according to the idea associated with it, i.e. in a wall-paper, foliage was called green, but a bow of ribbon in another pattern, though of precisely the same colour, was called pink.

The blue-blind are also uncertain in their names for the blue-green part of the spectrum. They find no change in the indigo near G, but their spectrum is not shortened. Their luminosity curve appears in some cases to have an appreciable minimum a little beyond F.

The violet-blind have the spectrum shortened in the violet and see no change of colour near G in the indigo.

If a colour sensation is weaker or stronger than usual it will occupy the normal position in the spectrum, but will extend a shorter or a longer distance on each side of its maximum. Thus, if the 'best yellow' is found in what we consider orange, it may indicate either a weak red or a strong green. If the former, the spectrum will be somewhat shortened in the red. If the latter, the red will be of full length, but the blue-green junction will be well beyond F. I have found several cases in which the green sensation predominated to such an extent as to render the vision practically monochromatic.

It is necessary always to allow time for the eyes to recover from the effects of any unusual fatigue on the day of the examination. I have known a man give all the signs of partial green-blindness after watching a cricket match in the hot sun. When this is the case, a second measurement of the red-green and green-blue junctions after half an hour will show a tendency to return to the normal, which may be taken to indicate that the condition is probably merely temporary.

It should also be noted that the novelty of the experience renders it a considerable mental strain to some people, so that it is well to get the more important measurements made as soon as possible. If, therefore, the colour-boundaries are approximately normal, proceed with the fatigue in the usual way. But if the first striking change of colour occurs in the middle of the green, test at once for a red-blue overlap.

In a case of partial green-blindness the red-green and green-blue overlaps may come so near to each other as to leave very little free space between. Under these conditions, although on fatigue with red the red-green boundary moves towards the red, the reaction for green cannot be obtained, there not being enough of that colour to fatigue with, but on making the attempt the region of colour-change moves up quickly as far as b. This is an almost certain sign of a very weak green.

Similar phenomena occur between F and G with a very weak blue sensation, and this part of the spectrum is described as being of a nondescript or neutral colour, or some kind of blue with a contemptuous epithet—bad, poor, dirty, mangy, &c.

To sum up. Deficiency in the red or in the violet

sensations is indicated by a shortening of the spectrum. at the red or at the violet end. Insufficient green affects the reactions between D and F, and a weak blue sensation causes similar confusion between F and G. But when cases of this kind are met with it is generally better to ask for a second interview, and having decided from careful consideration of the first results exactly what portions of the spectrum are to be used for producing fatigue, to spend all the time upon them.

*V. 4. Colour-change and Wave-length.

There are various ways of investigating this subject. For that here described the apparatus used is a grating-spectroscope by Beck having an eyepiece furnished with an adjustable diaphragm or sliding shutter wherewith to regulate the width of the field of view.

The diaphragm is opened till the observer can just detect a difference of colour between the two sides of the part of the spectrum visible through it.

In some parts of the spectrum the colour-change is very rapid, and in others the colour is constant for a considerable distance. The object is to draw a curve expressing these variations of the rate of change of colour *as it appears to your own eye*. For this is one of the 'personal characteristics' of colour sensation depending, as all such characteristics do, primarily on the individual, but also on his physiological condition, on the intensity of the light used, and on the time spent on the observations.

It is important, therefore, in order to ensure a reasonable regularity in the results, to work at a steady rate, to keep the illumination constant, and to complete the entire set of observations at one sitting.

The results should be plotted on the same curve as those of the Measurement of the Colour Sensations by Burch's Method, and compared with them.

It is most instructive to measure the change of colour sensations on a grating-spectroscope and the colour sensations themselves on a prism-spectroscope. In spite of the great divergence of the scale-readings, the results, when reduced to wave-lengths, tally in a remarkable manner.

N.B.—It is possible to detect and to investigate colour-blindness by studying the relation between colour-change and wave-length.

1. To find the correction for the scale-readings of the eyepiece, and to calibrate them.

The left-hand edge of the diaphragm shutter is fixed, the right-hand edge being actuated by a screw of 1 mm. pitch, with a head divided into 100 parts. The width of the aperture is, therefore, measured to the hundredth part of a millimetre by the scale-readings of the eyepiece—we have to find the equivalent of this in wave-lengths. The position in the spectrum of the aperture is given by the spectroscope scale-reading referred to the fixed left-hand edge of the diaphragm—and we have to find the difference in wave-lengths between this and the cross-wires in the middle of the field which were used in obtaining the calibration curve.

We may, therefore, set up a sodium flame and read the spectroscope scale when the right-hand line touches the left-hand edge of the diaphragm. The difference between this and the reading for the cross-wires of the ordinary eyepiece must be added to all the spectroscope readings, and it will convert them into scale-readings of the calibration curve.

The question may be asked, will this correction be the same for all parts of the spectrum ?

To determine this point, wait for sunlight, and compare readings of the lines *B, C, D, F,* and *G* for the edge of the diaphragm with those obtained with the cross-wires of the other eyepiece.   ·

2. To calibrate the scale-readings of the diaphragm eyepiece, open the aperture to a width of exactly 1 mm. and take the difference between the spectroscope scale-readings when the sodium line touches first one edge and then the other of the aperture.

If the calibration curve previously obtained is truly a straight line, there will be this same difference of scale-readings for the width of an aperture of 1 mm. in all parts of the spectrum.   With the type of instrument used the difference may generally be neglected.   But it is as well to ascertain whether this is so by taking readings for *B, F,* and *G.*

By means, then, of the calibration curve the scale-readings for width of diaphragm can be expressed as differences of wave-length—generally by means of a constant multiplier.

3. Proceed to observe the changes of tint as follows :— Use by preference a Nernst lamp, as being steadier and more free from bright lines than the arc, and brighter than the glow lamp.   Focus the filament on the slit by a 3-inch lens of say 8-inch focus, the slit of the spectroscope being about 3 feet from the lamp and 2 feet from the lens.   A sheet of cardboard with

a rectangular slit in it ¾ inch long and ¼ inch wide should be supported in an adjustable stand just in front. of the spectroscope so as to completely shield the observer from the glare of the lamp.

Some trouble will probably be experienced from the accidental turning of the focusing screw instead of the screw regulating the width of the diaphragm aperture. To prevent. this after the instrument has been focused once for all, a disk of brown paper may be folded over the milled head and tied as a cap is tied over a cork in a bottle—but loosely. It may be fitted over a larger milled head first.

Then set the spectroscope to the beginning of the red. Open the diaphragm till you can see a difference in the colour of the right and left sides of it. In this case the difference will be of intensity. Record .the width and set the spectroscope to the next convenient graduation. It will be found necessary to take readings for every 25th degree of the spectroscope scale where the change is rapid and every 50th at other parts. For some distance in the red the colour is so far. constant that intervals of 100 divisions would be sufficient. More is, however, gained by the additional practice in rapid estimation than would be saved if the intermediate 50th divisions were omitted. For in this case, as with all such personal measurements, the best results are got when the observer works at a steady pace throughout—not hurrying, and not delaying to get too careful a reading in any one place.

As soon as the C line is passed the change becomes, more rapid, this time in hue. as well as in brightness— the green sensation mingling with the red. If it should seem that the most rapid change of all occurs between two of the positions taken an additional observation should be interpolated.

With normal colour sensations another rapid change occurs between blue and green, and a third less definite between violet and blue.

4. Plot your results, taking wave-lengths as abscissae, and widths of diaphragm as ordinates. If these are taken so that 1 mm. is plotted to the same size on the ordinates as 400 Å. U. on the abscissa, the curve will be well proportioned.

This is the method of plotting employed by Helmholtz in discussing the results of König, Brodhun, Uhthoff, and others.

Like so many of the observers from 1870–90, they do not seem to have carried the investigation beyond the blue.

Helmholtz gives two minima, one near D in the yellow-orange, and the other near F between blue and green.

Working on similar lines quite independently in 1889, I got both these and a third minimum near G between blue and violet.

All who are sensitive to violet as well as blue find this third minimum.

Each minimum marks the point where two adjacent colour sensations overlap.

By plotting this curve on the same sheet and to the same base line of wave-lengths as the extent of your colour sensations ascertained by Burch's Method, you will see how very closely the two are related. There is room for both along with the calibration curve.

The curve given in the upper part of Fig. 13 is that of a typical case of green-blindness. There is no minimum in the yellow near D, but a very marked one in the middle of the green at E between red and blue. There is no minimum between blue and green at F, but a well-marked one between blue and violet at G. There are therefore no signs of green, but the colour sensations are red, blue, and violet.

*V. 5. Edridge-Green's Monochromatic Colour-patch Method.

Edridge-Green (*Proc. Roy. Soc.*, vol. 82 B, p. 458) classifies people according to the number of monochromatic colour-patches they can distinguish in the spectrum, and has devised a special spectrometer for the purpose. In this instrument, made by Adam Hilger, two adjustable shutters with vertical edges moving in the focal planes of the telescope take the place of the adjustable diaphragm of the eyepiece in the instrument described in the previous section. For greater convenience the whole is graduated in wave-lengths, so that there are no calculations to make.

The mode of procedure is, however, different. No notice is taken of changes of luminosity, only changes of hue being considered. The right-hand shutter being brought to the extreme end of the red, the left-hand shutter is screwed back until a part is reached where a change of hue is just discernible. (N.B.—A prism being used, the red lies to the right-hand, whereas in the grating spectroscope it was on the left.)

The right-hand shutter is then moved up to it and the

left-hand shutter screwed out again until a change of hue is again visible.

And so on ; stepping through the spectrum with long or short steps according as the colour changes slowly or rapidly. Some people with this instrument make as many as twenty-nine steps, others only half a dozen or fewer.

Obviously the results are capable of being plotted in the same way as those of the previous section. The short steps correspond with the minima of those curves, the main difference being that in them readings are taken at equal distances throughout the spectrum even where the colour-change is slight, so that a better shaped curve is obtained.

Moreover, the number of monochromatic patches can be read off from the curve of V. 4 as follows :— Suppose for simplicity's sake that a slit 1 mm. wide takes in a difference of wave-length of 100 Å. U. At $\lambda 6500$ the slit width as shown by the curve was 2·5 mm., i.e. 250 Å. U. This makes the next step $\lambda 62500$, at which point the slit width = 2·1 mm. Therefore the third step will be $\lambda 6040$, where the slit width is 1·15. This brings us to $\lambda 5925$ and so on.

Of the two methods, that first described gives most information to the eye. Moreover, I fail to see any valid reason for excluding the changes of luminosity, since the changes of colour are merely the result of an increase in the luminosity of one constituent and a diminution of the other.

One very remarkable fact may be noted—that in spite of the great difference in dispersion of the two instruments, the general results show good agreement.

The more powerful grating spectroscope may be used in the way described by Edridge-Green and the preceding statements verified.

*V. 6. To Map the Field of Vision for different Colours with the Perimeter.

Perimeters are of various types, some requiring a second person to plot the observation and others being self-recording. That of Adams is both convenient and simple.

1. Place a blank chart in the frame, fitting the central pin exactly to the centre of the circle for the right or left eye, as the case may be.

See that the carrier on the movable quadrant runs easily. It is well to place a small slice of cork on the pin of the pricker, so that when pressed home it may mark the chart without tearing it, as it will if the point projects too far. Keep the frame clear of the moving parts except when making a record.

Put a small patch, e.g. of red, on the carrier.

2. Keeping the head against the rest, let the observer look steadily at the fixation point while the operator brings the carrier nearer and nearer to the line of sight. As soon as he perceives the *colour* of the patch, make the record by pressing the frame carrying the chart against the recording point. The patch will be visible before its colour is perceived.

If preferred, readings may be taken of the points at which the colour disappears as the spot is moved outwards. Readings should be taken in each of the directions marked on the chart.

White, red, green, and blue spots should be used, bearing in mind, however, that none of these is even approximately a pure spectral colour, except perhaps the red.

The colour fields of each eye must be taken.

*V. 7. Dark Adaptation.

Experiments on dark adaptation differ so much in detail according to the apparatus available and the particular arrangements of each laboratory that it would be futile to attempt more than the indication of general principles.

For all quantitative work the essential condition is that the darkness should be complete. That is to say, it must not be possible, after the longest stay in the dark-room, to detect any crack, or pinhole, nor to perceive any indication of the presence of light save what is required for the purposes of the experiment. If this condition is not fulfilled the measurements will be merely relative, and will correspond to a much less complete condition of adaptation than should have been reached in the time.

Experiments with coloured pigments or coloured films or glasses have a merely historic interest. They are quite unreliable and lead in many cases to erroneous conclusions, owing to the complex character of the spectrum of all but a very few pigments.

Inasmuch as dark adaptation comes on at a different rate for each colour sensation, the only really satisfactory plan is to use spectral colours. This may be done in two ways, viz. by projection, or by direct vision.

For the *Adaptometer by Projection* we may arrange an ordinary spectroscope, preferably of the constant deviation type, to project the spectrum on a screen. By means of a diaphragm all but the particular colour to be studied is hidden. The resulting colour-patch is used as a source of light to illuminate a second white surface at which the observer looks. The intensity of the illumination can be varied in two ways, viz. by altering the distance between the colour-patch and the second screen, and by altering the area of the colour-patch. By a combination of the two it is easy to get a range of from 1 to $\frac{1}{1000}$th part. The whole is of course enclosed in a blackened case.

For the *Adaptometer by Direct Vision* a spectroscope such as that described in Section V. 4 may be employed, the diaphragm of the eyepiece being set so that the portion of the spectrum visible is practically of the same colour at both edges, without being too narrow.

The intensity of the light must be reduced before it enters the spectroscope, otherwise there will be no possibility of getting the ends of the spectrum pure. A camera-body or box about two feet long is fixed against the shutter of the dark-room. At the end of this is placed a white screen—card or porcelain, or other dead-white surface—at $45°$, to reflect the light into the spectroscope, the collimator tube of which passes into the box from the side, close to it.

This joint must be absolutely light-tight. To make it so, cut holes the exact size of the collimator tube in two black cards six inches square, cut these in half the one vertically, the other horizontally—fit them together round the tube behind the flange of the slit frame, with a washer of black cloth if preferred, and clamp the whole firmly against the side of the box.

N.B. Clamps and thumb-screws should always be used when an instrument has to be fixed to some piece of apparatus, so that it carr be removed for other purposes with the least possible trouble. Each clamp should have a hole close to one end through which one of the thumb-screws passes. Near the other should be a slot cut in from one side to just beyond the middle. The thumb-screws are merely slackened and each clamp turned aside in order to remove the instrument. In this way there is nothing that can get lost, for a black thumb-screw in a blackened dark-room is not easy to find.

With this arrangement there is no visible 'field of view'. Under ordinary circumstances the spectrum can be seen as a band of light stretching across the circular area of the eyepiece diaphragm. That is owing to 'false light' reflected into the instrument by the jaws of the slit, or falling on the slit at a greater angle than the collimator can take up, or reflected from the edges of a lens or prism. Frequently it is due to light from one or other of the ends of the spectrum falling on the sides of the eyepiece tube. Instrument-makers will persist in acting as though a blackened tube were the theoretical 'black body'.

No part whatever of the eyepiece diaphragm should be visible save where the spectrum crosses it.

Intensity of Illumination.

It will be observed that the width of the slit remains unaltered through the entire series of experiments which it is intended to compare. Measurements of slit-width are not sufficiently reliable for work requiring an accuracy within one per cent. over a range of 2000 to one.

If daylight is used, the easiest way of regulating the intensity is to vary the size of the aperture in the shutter of the dark-room. But it must not be forgotten that daylight may vary very considerably both in composition and in quantity during an experiment, so that photometric observations must be made to correct the results by.

With artificial light the simplest plan is to vary the distance between the lamp and the white card screen. For still greater reduction Ayrton's method of interposing a short-focus lens, and so forming an image of the flame, reduced in size, may be used. Or a surface of unsilvered glass may be employed to reflect the light on to the screen, the ratio of reduction being separately determined once for all, by a photometer. N.B.—A plate gives reflections from two surfaces, a prism from one.

To obtain the curve of increased sensitiveness to a given colour during dark-adaptation. Set the spectroscope so that only the colour selected can be seen through the diaphragm of the eyepiece. Note the time at which you enter the dark-room. The illumination having been reduced below the visible limit, let it be increased until you can just perceive light, and then further increased until you can distinguish colour. Let both readings, and the time, be taken by a colleague.

Let the light be again reduced below the limit and wait ·five minutes. Then look again, and repeat the previous operations. Do this every five minutes, recording the time, the intensity of the light when just visible, and the intensity when colour can be distinguished. These two intensities are very different at first, but they become less so as the after-effects of previous illumination die out, and finally, in my own case, the difference disappears when dark adaptation is complete. This may take two hours or more.

Remember that each colour sensation extends much farther in the dark adapted state. Pure red will be found at B. Near C the spectrum will excite a marked proportion of green. The nearest approach to green occurs near D—at E it is largely mixed with blue. Pure blue is not met with anywhere because of the extraordinary increase in sensitiveness to violet, but pure violet may easily be seen at H and K.

Relative Visibility of Two different Colours.

In my paper on Colour Vision by Very Weak Light[1] I have described a method of measuring the relative visibility of two different colours by direct comparison. In this case the instrumental readings are recorded in the dark by the observer himself, and afterwards evaluated by daylight. For details the paper should be consulted.

It may be mentioned that for taking notes in the dark it is best to use a reporter's notebook, opening endways, with pages about 4 in. × 7 in. This is held in the left hand with the four finger-tips projecting well above the paper. If the little finger of the right hand rests against each of the four fingers in succession it is quite easy to

[1] *Roy. Soc. Proc.*, vol. 76 B, p. 199.

write a line, curved it is true, but quite clear from the other lines. Number each note before you begin it. *Write rapidly* and keep the pencil on the paper while you are thinking. Do not attempt more than four lines to a page at first and when each page is full, tuck it under a rubber band and begin the next note on a fresh page. The time to which each note relates should be recorded by a signal mark on a cylinder covered with smoked paper and driven by clockwork, or some equivalent arrangement.

It is amusing, especially for the first two or three occasions, for the observer to note down his own estimate of the times that have elapsed during experiments extending over, say an hour.

When working alone especially at night, a clock striking the quarters, or better, a repeater, is a great convenience.

Fixation Points.

It is frequently necessary to have in the dark-room one or more extremely minute points of light to guide the eye in the required direction. Especially is this the case with experiments on peripheral vision by flashlight.

Such points must be equally visible over a fairly wide angle, otherwise an unconscious change of position might cause the observer to lose them. Moreover, there are many experiments which can be demonstrated to a group of people sitting together. They should be portable, and capable of a considerable range of adjustment as to brilliancy.

The simplest kind consists of a glass rod, the size of

a pencil, cemented with plaster of Paris in the side of a box containing a candle or lamp. The end inside the box is fused to a hemispherical shape so that it may collect the light like a bull's-eye.

The other end is drawn out to the thickness of a knitting-needle and the tip of it fused into a bead. Light entering the large end is totally reflected along the sides until it emerges by the bead at the small end. The rod need not be straight, but may be curved so as to point in any direction. The issuing light is slightly coloured green or pink or brown, according to the kind of glass used. The intensity is regulated by the distance of the lamp from the large end.

A better form consists of a brass tube, blackened, with a cap at each end, in which is mounted a spherical glass bead—or a lens of short focus. This must be fixed in the side of the lamp-box facing the observer, or where that is not convenient, a diagonal reflector, either between the two lenses or beyond the second, may be used to throw the rays in the required direction.

If the presence in the dark-room of a small amount of light is not objected to, the fixation spot may be the reflection in a convex mirror of a screened lamp near the observer. There are some advantages in this plan. It is convenient when changes have to be made in the colour or character of the light used.

Fixation Points with Spectral Colours.

For some purposes it is essential to employ spectral colours. In such cases a spectroscope with diaphragm eyepiece, such as that used for studying Colour-change and Wave-length by Helmholtz's method (Section V. 4), will answer admirably. Or in place of the eyepiece a cap containing a spherical glass bead will serve. Owing to the small diameter of the bead no diaphragm is necessary. A cap containing a row of beads at suitable distances will give a row of points of different colours, or of the same colour, according to the direction of the row with regard to the spectrum.

In place of a spherical bead a piece of glass rod one or two millimetres in diameter and two or three centimetres long may be substituted for the eyepiece. This will give, when set parallel to the slit, a very fine line of monochromatic light of any desired wave-length.

It is of course understood that all such arrangements are intended to be looked *at* rather than *through*; i.e. the observer should be at arm's length or more from them. They are visible over a fairly large angle, but give maximum illumination in the direct line of the ray.

*V. 8. The Limits of the Visible Spectrum.

To observe the limits of the visible spectrum is a different problem entirely from that of determining the faintest visible light. For it is necessary to receive into the spectroscope the most powerful light attainable and yet to exclude all false light from the eyepiece.

Since the rays are parallel after passing the collimator, the utmost it can take in is given by the angle subtended by the lens at the slit. The whole of this, however, is seldom utilized, the makers being more chary of prisms than of lenses, and the light so admitted falls on the rough edges of the prisms or the sides of the tubes, and constitutes a considerable percentage of 'false light'.

It is a very good plan to take out the objective of the telescope, substitute a cork with a hole in it for the eye-piece, and looking through this to ascertain, by pushing a piece of card between the collimator lens and the prism, exactly how much of it comes into use in forming the spectrum. Then make a diaphragm that will stop off all the rest. Place another similar diaphragm on the near side of the prism.

Finally, and this though most important is very seldom done, cut a diaphragm of the same shape as that by the collimator, but exactly half the size, fit it to a black tube six-tenths the length of the collimator-tube, and fix it *in front of the slit*. Focus the light on the slit by a lens placed outside this tube at any convenient distance.

The angular aperture being limited to something less than that utilized by the collimator, the beam will not even touch the diaphragms inside the spectroscope.

There will still be stray light in addition to that due to dust, scratches, bubbles, or imperfect polish, and for this reason. At every surface by which light is refracted, a certain percentage of it also is reflected. Thus at the first surface of the prism there is some reflection, and again at the second surface. In this case a little consideration will show that the light must be reflected back on to the base of the prism, and thence again, if this is polished, to the first face, and thence to the second from which some of it will emerge in the form of a *white image of the slit*, easily detected if an equilateral prism with the three faces polished is put on the goniometer.

For spectroscopes the third face is not polished, and the rays thus reflected on to it must be absorbed by a black varnish made by mixing lamp-black with a dark-red varnish, such as is used by photographers—the red being used because lamp-black reflects an appreciable amount of the more refrangible rays.

A great deal of false light is due to the reflection of some other part of the spectrum from the inside of the tubes. I have already remarked on the difficulty of persuading instrument-makers to make these big enough.

If the telescope is placed some distance away from the prism instead of being close to it, the dispersion is not altered, but the field of view is limited. In this way it is possible to get rid of a good deal of the false light reflected from other parts of the spectrum. Another plan is to use two spectroscopes in tandem.

In addition to all these precautions coloured screens may be employed to cut off all but the extreme red in the one case, and the extreme violet in the other. To hold liquids transparent to ultra-violet rays, vessels of Uviol glass may be used, or tanks made of 'axis-cut pebbles', which may be obtained in plates about $1\frac{1}{2}$ inches square quite cheaply of the spectacle-makers.

It goes without saying that the observations should be made in the dark-room. Full sensitiveness to ultra violet is attained only after a stay of at least two hours in absolute darkness. The prism and lenses of the spectroscope should be of quartz or calcite specially constructed for the purpose. The preliminary arrange-ment of the instrument—focusing, light, &c.—must be done by some one else.

Diffraction gratings are not suitable for this work. Even the spectrum of the first order is confused at the red end with ultra violet from the second, and at the violet end it is mixed with white light diffused from the direct image of the slit.

V. 9. Confusion Tests for Colour-Blindness.

Nagel's.—A series of cards, each bearing a ring of spots the size of peas, of various shades of pale red, green, brown, crimson or purple arranged so as to con- fuse the colour-blind. Full directions are given with each set of cards, which should be kept away from the light when not in use.

Stilling's.—A series of cards covered with irregularly shaped spots in colours likely to be confused. Thus for red-blindness a number or date is written in red spots and the space filled in with spots of green or brown, and still farther to confuse the eye, a number of false trails are led in all directions in spots slightly darker than the rest. Similar cards in different colours are employed for other varieties of colour-blindness. Thus, for instance, the green-blind confuse various shades of green and purple. Some of the designs are difficult to trace, even to the normal sighted. These are used for the detection of cases of slight colour-blindness.

Burch's.—This consists· of a set of cards painted in colours which appeared exactly alike to the late Lt.-Col. W. R. L. Scott, who was red-blind. Thus Vermilion is matched by Green the colour of a year-old ivy leaf. On the green half of the card are painted the words DONT GO, the letters NT being in vermilion and the rest in blue. He could see the blue but not the red, save by holding the card sideways so that the light fell on the brush marks. To render these invisible a sheet of perforated zinc is fixed some 6 inches above the card, and the observer looks through a tube containing a convex lens of 8 diopters focused on the zinc. The card is so far beyond the focal plane that all brush marks are invisible, and given the colours any design can be painted in a few minutes.

The other cards are Geranium Red, matched with a French Grey or bluish slate-colour, Emerald Green, matched with a rich ochre made largely with Mars yellow, and Lilac, made with mauve, magenta, and white, matched with a pure blue. The red-blind are unable to see geranium-red letters against French grey, or yellow ochre letters against green. But they instantly perceive salmon-pink against green, though the difference to normal eyes is very slight. They cannot distinguish blue letters on a lilac ground, but they can easily see letters of a slightly paler blue upon the blue itself, although not one in twenty with normal sight can do so. Green-blind people can see the lettering on all the cards save the last, the blue and lilac appearing alike to them.

Ryland's.—This confusion test consists of a pair of disks of ordinary green glass which appear of a darker green when superposed, and a second pair of special kind, each of which singly is green, but the two when superposed look red.

The explanation is that certain films and liquids which are partly transparent to green are extremely transparent to a narrow band in the red. Thus in thin layers the green transmitted overpowers the red, but in thicker layers so much of the green is stopped that the red asserts itself. Now although the red-blind man is as likely as not to say that the two disks when superposed look red, he will say the same also of the disks of ordinary green, for to him dark green and bright red are alike. But the normal sighted can generally be recognized by their surprise at the change if they have not seen it before. It is necessary to suit the light to the particular disks employed.

For some, a candle will answer, for others an incandescent gas-lamp. Some will work by daylight—others do best with a glow-lamp.

Cobalt glass looks blue when of moderate thickness, but a glow-lamp seen through a thick piece appears red.

Permanganate of potash on the contrary is pink in weak solutions and deep violet in strong ones.

V. 10.   The Holmgren Wools Test.

The method here given describes my own practice, and has no official authority. It differs in some respects from the Board of Trade Regulations which came into force on January 1, 1910, and must not therefore be substituted when these are prescribed.

1. Turn out the wools in a loose heap on some *dark surface, preferably black*. A white cloth makes the dirt on the skeins show up, and the colours can be more easily distinguished against something dark.

2. Pick up a skein of a deep bright red, free from any tinge of crimson on the one hand, or of orange on the other, and ask the person tested to match it. I use a set containing *at least one perfect match for each of the important colours*, and I ask for all the skeins that match to be found. Watch the process of selection—whether instant or hesitating, but bear in mind that some people are nervous and others inexperienced in remembering the shade they want to match until the two are side by side.

If dark greens and browns are gathered with the bright red, there is probably red-blindness—if orange and bright greens are picked up, green-blindness is indicated.

3. A rather blue purple, in colour between blue serge and purple clematis, is matched by the red-blind with brilliant magenta, but by the green-blind with green.

4. After a while you may pick up a dark green to be matched. The red-blind will put it with scarlet, and the green-blind with slate colour. If many skeins are handed in as matching the sample I ask which is considered the best match, and desire to have them arranged in order of nearness to the sample colour. In this way it sometimes appears that the notion of matching a given colour has never been grasped. It is important to remember that one is testing the *sensations* and not the education of the candidate.

Violet-blindness may be recognized by the person tested confusing the old-fashioned dull purples with the modern aniline purples and violets, and there is a certain very intense and brilliant aniline violet which they place with black or dark red-brown.

Confusion colours should be studied by means of the apparatus used by Helmholtz for combining colours, which consists of a horizontal stand carrying at its centre a vertical mirror of unsilvered glass. Take a pair of colours which match to the red-blind, viz. bright red and dark green. Lay them side by side on the stand behind the glass, and in front of the glass lay a sheet of white paper. Stand so that the reflection of the white paper covers the skeins, the colour of which will appear diluted by the white. Pick out from among the other skeins some which are of the same tints as the colours thus diluted. By arranging the light so that it may be strong on the white paper and weaker on the skeins, you can make the tints so pale that the difference is barely perceptible. These are all confusion-tests of more or less difficulty.

For the green-blind let your colours be green and purple.

Instead of white you may use brown or green or bronze. You will then have a series of dark confusion colours which are just as efficient in detecting partial colour-blindness as the pale ones, with this advantage — that whereas many of the uneducated will tell you that they 'don't reckon to be a judge of fashionable colours', they will attack these darker ones with confidence.

During the examination of a case of colour-blindness the same etiquette must be observed as in any other kind of diagnosis—or by the spectators at a game of chess.

V. 11.   Edridge-Green's Signal Lantern.

This is intended for testing railwaymen and seamen for colour-blindness. The light of a lamp is reflected through an aperture, the size of which is regulated by diaphragms to represent a signal lamp at various distances. Coloured glasses of different shades of red, green, blue, &c., can be brought into play, or the light may be dulled by ground glass without changing its colour.

Full directions are supplied with the instrument by the makers.

SECTION VI

EXPERIMENTS BY FLASHING LIGHT

VI. 1. Composition of Colours.

Newton showed that certain colours combine to form white, by spinning a disk painted with the colours in sectors of the required proportions.

Helmholtz made it possible to adjust the relative proportions of the colours used by employing disks of coloured paper each slit from circumference to centre.

To use these, first place upon the axle of the whirling apparatus a metal or card disk as big as the biggest of the paper disks. Then put on the paper disks, passing each through the slit of the one behind it, so that part of it is hidden and part visible. The relative proportion of the colours is adjusted by pushing the visible part farther under, or drawing it out, as may be required. When this has been done, put on the metal washer, and screw the nut over it to prevent shifting.

N.B.—The free edges must be so directed as to *follow* the rotation, otherwise the papers will be torn.

1. A certain orange-yellow, combined with blue in proper proportions, will make a very fair white. Do this with 6-inch disks.

2. Red and green and a certain blue-violet in proper proportions, will give an equally good impression of

white. Do this with 4-inch disks on the top of the other. You will then be able to compare the tone of one mixture with that of the other.

3. The whites so obtained might be more truly called grey. Take off the orange-yellow and blue combination and substitute for it a pair of disks one black and one white. Match this with the red, green, and violet mixture of the 4-inch disks. You will thus obtain a measure of the 'luminosity' of the mixture, i. e. the percentage of white to black in the compound white you have made.

Note that—I. The true colour of the mixture is only reached when the disks are rotating with great rapidity.

II. The component colours appear of extraordinary brilliance when the disks are going very slowly, and look much duller when they have stopped.

VI. 2. Verification of Talbot's Law.

Set up the photometer—grease spot, paraffin block or Trotter photometer, and interpose the revolving sectors in the path of the light on the side of the fixed lamp. Set the distance for 50 cms., i. e. place the movable lamp at 50 cms., adjust the other lamp till they balance, and let it remain at that distance.

Then start the motor at such a speed that flickering ceases and the light appears steady. Set the movable lamp in succession at 60, 70, 80, 100, 120, 150, 200 cms., and adjust the sectors until the illumination of both sides is equal in each case.

According to Talbot's Law, if a = duration of flash and b = duration of eclipse, then $\dfrac{a}{a+b}$ = intensity of illumination given by the flashing light, taking the steady light as unity.

This law holds so long as the flashes are of such frequency that 'flicker' is no longer visible.

If cardboard sectors are used it is better to set them to a known flash ratio, and adjust the movable light until the two balance, because in this case the flash ratio cannot be altered while the machine is running.

The rate at which flicker ceases should be noticed.

Experiments should be made also at slower speeds and the rate determined at which Talbot's Law begins to fail.

VI. 3. The Failure of Talbot's Law.

A disk of black card has three concentric rings of sectors cut away.

In the outer, and also in the inner ring, there are eight of these apertures equally spaced and of equal angular magnitude.

In the middle ring are only two apertures each subtending double the angle.

Thus the flash ratio of the inner and outer rings is double that of the middle ring, but the frequency is also twice as great. If this disk is revolved slowly in front of a sheet of white paper, or a mirror reflecting the light of the sky, the middle ring looks brightest, but if it is revolved rapidly the outer and inner rings look brightest.

Find at what speed both look equally bright.

*VI. 4. Photometry of Flashing Light.

Set up the Episkotiston (or Adjustable Revolving Sectors) on one side of the photometer head in the path of the light. Either a paraffin block, or a Trotter photometer, will serve equally well.

Instead of the Episkotiston, a cardboard disk with fixed sector will do, but it should be loaded with the brass disks to make it drive steadily.

The electric motor should be used, geared down so that a speed of from one to ten flashes per second may be within the range of the adjustments.

Suppose the sectors to be to the right of the photometer head.

With the sectors stopped, adjust the right-hand lamp to balance the left-hand lamp when fixed at 100 cms.

Then having started the motor, adjust the left-hand lamp until *each flash appears of the same brightness as the continuous light.* Keeping the number of flashes per second constant, vary the aperture of the sectors from 10° to 20°, 40°, 60°, 90°, taking readings for each.

Then repeat with a different flash rate, taking a set of observations at each of the speeds 1, 2, 5, 7, 10 flashes per second.

If the cardboard disks are used, as the aperture of the sectors cannot be changed without stopping the motor, it is easier to vary the flashes and take observations for all the flash rates before changing the sector.

Plot the results, remembering that the intensity of the light varies inversely as the square of the distance.

VI. 5. To Adjust the Heliostat.

1. If the clock does not start ticking on winding it up, a gentle rotary shake about the axis of the hour hand will make it do so.

2. Having taken off the mirror put the sighting bar on the hour axis, and set it to the time of day.

3. Stand the heliostat in the sunshine, on a level surface in the position it is to occupy, and adjust it so that sunlight passing through the pinhole in one arm of the bar falls on the day of the month marked on the other arm. Then tighten the set-screw.

4. Remove the sighting bar carefully without disturbing the heliostat, and put the mirror in its place.

5. The mirror may be set to any convenient position on the hour axis, but the angle of incidence of the light should not be made too great. If the laboratory is so situated that the reflected beam has a suitable direction when horizontal, the spectroscope may be adjusted to that line. If not, let the heliostat mirror reflect light on to the second mirror, by which the beam may be reflected in any required direction. There is a certain loss of light in so doing, but what is left is ample for nearly all purposes. There is, moreover, this great advantage, that after the apparatus has been set up, the light can be adjusted by merely moving the mirror.

If well set, the heliostat should not need readjustment for three or four hours. If after keeping steady for a while the beam begins to move steadily, it is generally a sign that the clock has stopped.

VI. 6. Experiments with Flashing Light.

Experiments with flashing light may be divided into two groups, namely, those which demonstrate a phenomenon, and those by which the cause of it may be investigated.

To the first class belong the great majority of experiments with coloured disks, and to the second those made with the spectroscope. The first are by no means to be despised since by their means most of the phenomena have been discovered. But as from the very nature of these observations time-relations are involved, and as the time-relations of the colour sensations are different, it is practically impossible to disentangle the phenomena when such mixed colours are used as those of ordinary pigments.

Among the observations that may be made with the ordinary Helmholtz Whirling Table are the following :—

Rate at which flicker ceases for—

Black on white,   red on white,   green on white,
Blue on white,   red on blue,   red on green.

Rate at which red and green look most brilliant.

All experiments on Areal Induction and Shelford Bidwell's experiments are worth doing in this way.

The methods available for determining the number of alternations per second are simpler with hand-driven than with motor-driven apparatus.

Turn the apparatus slowly and ascertain how many revolutions of the disk correspond to one of the handle. Then use a metronome set to some suitable rate and keep pace with it turning as steadily as possible. If the rate of the handle is too slow for the range of the metronome put a white or red paper mark on one of the intermediate multiplying wheels and keep pace with that.

VI. 7. The Spectrum by Flashing Light.

It is essential for the success of the experiments described under this head, that the flashes should be regular in duration and occur at regular intervals, and that the beginning and the end of each should be sudden. If it is immaterial—as in Benham's spectrum top and Shelford Bidwell's experiments and the like—whether the change occurs all over the field at once, or not, then the effect of a flash may be got by a disk with slits revolving in front of the object, or in the case of the spectrum, in the focal plane of the eyepiece.

But if it is essential that the flash should appear all over the field at once, the light itself must be made intermittent. Owing to the construction of most spectroscopes it is not possible to get a revolving disk very near to the jaws of the slit, and the loss of suddenness in flash and eclipse consequent on even this small distance is very marked. Moreover, there is a certain risk to the instrument in causing a disk of 30 cms. diameter to make fifty revolutions per second close to the slit of a spectroscope not permanently part of the same apparatus. For this reason I prefer the arrangement shown in the diagram.

With the lens B an image of the source of light A is thrown upon the disk C. When the disk is turned, the light which passes through the apertures in it is again brought to a focus by the lens D on the slit

S of the spectroscope.   The final image must be big
enough to cover the slit from top to bottom, otherwise the
spectrum will not be equally bright all over the field
of the wide-angled eyepieces that must be used.

With the arc light, lenses of 3 or 4 inches focus may
be used, and the distances B-A and B-C being equal,
D-S must be two or three times as great as D-C, the
lens D being focused till the image is sharp.

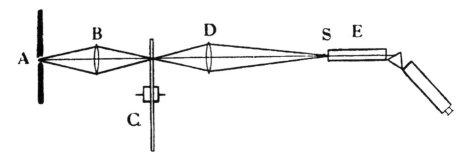

FIG. 14.

When all the adjustments are correct, on looking
through the spectroscope and slightly moving the disk
with the finger the light is seen to appear and disappear
all over the field with great suddenness.   A candle
should be substituted for the arc before doing this.
With sunlight a heliostat must be used to reflect the
rays horizontally in a fixed direction, and a lens of
longer focus—30 to 50 cms.—substituted for B in order
to get a big enough image on S.

It is dangerous, especially with sunlight, to look
through the spectroscope when arranged for these
experiments, except with the disk rotating.   It should
be driven at a good rate first, and allowed to slow down

while the observer watches until the desired effect is perceived.

The speed of the disk is regulated partly by the adjustable resistance in circuit with the motor, and partly, especially for slow speeds, by the soft leather break on the shaft of the disk. A heavy disk conduces to steady speeds.

And it may be necessary here to warn those who fit up their own apparatus that it is highly dangerous to run such a disk between centres. A sudden strain may cause the shaft to jump its bearings. I have seen such an accident happen. A 12-inch disk of sheet zinc revolving at a high speed shot across the room and struck the back of a Windsor chair, cutting clean through it like a saw. The bearings should be fairly long. The ends of the shaft should be pointed and should project beyond the bearings on either side, resting against steel thrust plates furnished with set-screws to prevent any undue play end ways.

To determine the number of flashes per second at any given moment, the disk has a series of holes round its edge so as to act as a siren when a stream of air is directed on it through a tube fixed to the surrounding frame. The note given out is identified by comparison with a piano or other musical instrument, and the number of revolutions calculated from the pitch of it.

It is difficult without a little practice to identify the octave to which the siren-notes belong, owing to the great difference between their tone and that of a piano or a tuning-fork.

*VI. 8. The Rate of Maximum Flicker of Spectral Colours.

A spectroscope of wide dispersion should be used and the slit should be long enough to fill the field of view of the eyepiece, which should be of wide angle.

The general arrangement should be as already described. If the light is strong enough a flash ratio of 1 : 44 will answer well, i. e. two slits 180° apart each 4° wide. With a weaker light longer flashes may be used, but there are certain phenomena only visible with the short flashes.

Set the motor going rapidly, direct the spectroscope to the green, and then slacken speed with the regulator and brake until the phenomenon of maximum flicker is observed, when the rate should be taken by a colleague.

The operation should be repeated for the different regions of the spectrum, of which the more important are, for this purpose, the centres of the four colour sensations, and the three parts where they overlap, namely, the yellow near D, the blue-green near F, and the indigo near G.

It is well to get readings for all these regions with a moderately narrow slit, and then take the pure red and the pure violet over again with a wider slit to compensate in some degree for the feebler luminosity of the ends of the spectrum.

The results may be plotted with wave-lengths for abscissae and flashes per second for ordinates.

VI. 9. Phenomena with very intense Light.
Subjective Absorption Band.

Using a single sector 2° wide with the most intense light obtainable, at about twenty-three flashes per second, a curious subjective absorption band makes its appearance between *C* and *D*. It is of a distinctly brown colour and is curved towards the yellow in the neighbourhood of the fovea. The light on the red side of it is of an extraordinarily rich red mingled with beetle-green in the fovea. A similar bend in the border line between the two colour sensations occurs where green meets blue and again when blue meets violet, but these are very variable, the bend being away from the red end of the spectrum in some cases and towards it in others.

It appears to depend on partial colour-blindness of the fovea.[1]

VI. 10. Purkinje's After-images.

Purkinje's After-images and Charpentier's Bands are phenomena respectively of the positive and negative after-effects which occur only after sudden illumination.

They are best seen when the eye is in some degree dark-adapted.

Purkinje's After-images may be shown by securing a piece of charcoal with wire to the end of a bamboo, igniting it in a bunsen burner, and then whirling it steadily round and round in a large circle in the dark. The observer should look, not at the glowing charcoal, but at the centre of the circle described by it, and the operator should keep it moving steadily, but not too

[1] *Journal of Physiology*, vol. xxi, 1897, p. 426.

fast. After a few moments the observer will note that the light, instead of being drawn out by the 'persistence of vision' (another name for the positive after-effect) into a long trail, is broken up into patches so as to produce the appearance of two or three images following the light itself.

A more elegant and safer way of demonstrating the phenomenon indoors, is to use a glow-lamp with a long piece of flexible conductor attached to the bamboo.

The most scientific method is as follows :—

A right-angled reflecting prism is mounted in the end of a hollow shaft, so as to reflect the light coming along its axis to a second prism carried by an arm attached to the shaft and revolving with it. This second prism can be adjusted so as to reflect the light into an eye placed in a line with the axis. Monochromatic light of any desired wave-length is furnished by a spectroscope, the optic axis of which coincides with that of the revolving shaft. If the rays issuing from the spectroscope are parallel, the apparent diameter of the circle described by the light is independent of the distance of the eye from the instrument.

VI. 11. Charpentier's Bands.

Charpentier's bands may be shown to an entire audience by means of a projection lantern. All that is required is a semi-disk mounted on a slide, so that it can be rotated steadily at the rate of a revolution in one or two seconds. The audience should be instructed to keep their eyes fixed on its centre, and at the same time to watch what goes on by the receding edge of the semi-disk near the periphery, *without looking at it.*

They will see one or more dark bands following the edge and keeping at a distance from it, depending on the speed of rotation.

A pointer, which must be small and unobtrusive, may be so attached to the disk as to follow it at a definite angular distance. If then the rate is so regulated that the band appears to coincide with the pointer and the number of revolutions per minute taken, the time interval between the falling of light on the retina and the development of the band, can be calculated.

A large cardboard disk, say 50 cms. in diameter with one half painted black, answers equally well for a single observer. It must be brightly illuminated and all the surroundings must be dark. A very good plan is to use a hood for the eyes, like a stereoscope, but without lenses, fitted to the face and padded with black velvet, having at the bottom two holes just big enough to see the disk through. The observations can then be made in direct sunlight.

VI. 12. Experiments with Rotating Disks.

A number of experiments can be made with disks half black and half white. They may be divided into two classes, namely, those which simply show that the sudden exposure of a given part of the retina to brilliant light induces a sort of momentary blindness, and those which also show that this momentary blindness extends to some distance in the retina beyond the parts on which the light fell.

Sigmund Exner's Experiment is of the first type. It was made with an apparatus fully described by Helmholtz. After an interval of darkness, a small white disk appeared for a fraction of a second. This was followed by a much larger white disk, and this again by darkness. The curious result was that instead of looking white the small disk appeared dark against the larger one, in spite of the fact that more light had fallen on that part of the retina.

To show the Negative Image of a Glow-lamp.[1]

This may be done with a disk of Bristol-board 20 cms. in diameter, half of which is painted black with Indian ink, and the other half left white. At the junction of the black with the white, along a radius, a slit is cut with a knife. This slit must not be more than about half a millimetre wide and three, or at the most four, centimetres long. Its edges may be pressed smooth by passing the blade of a palette knife through it. A glow-lamp being set up behind the slit at a distance of half a metre or so, the front of the disk is brightly illuminated with a couple of lamps with white shades to concentrate the light upon it. If the disk is made to rotate so that first the white half, then the slit, and finally the black passes before the eye, the filament of the glow-lamp behind will be seen as a thread of light once during each revolution. But if the direction of rotation is reversed, so that the order is black card, slit, white card, then the glowing filament will look like a black thread against the white card.

For a less brilliant light a wider slit must be used. Thus for an ordinary candle flame a slit of three or four millimetres will be required.

[1] *Proc. Roy. Soc.*, vol. 66, p. 212.

Shelford Bidwell's Experiments.[1]

By using coloured objects instead of white, Shelford Bidwell obtained very striking results. The disks he finally recommended had a sector of 60° cut away and of the remainder 150° was black and 150° white, or rather grey, which he found preferable to white. He placed a lamp of 50 c. p. with ground glass bulb in front of the disk, and the light, passing through the sector, served also to illuminate the objects behind it.

With such a disk, rotating at the proper speed, red objects appear green and green objects red, if certain precautions are taken.

Briefly, the main action is this: the momentary blindness produced by a sudden flash affects only the colour that produced it. What is seen therefore is the complementary colour supplied by the white paper. Now the complementary to red is popularly supposed to be green, whereas to any person of normal sight it is really a Cambridge blue. Similarly the complementary to green is not red but purple. To reproduce the popular conception of contrast we must therefore *get rid of the violet.* This may be effected either by illuminating with rather old carbon glow-lamps or by employing yellowish paper instead of white. If this is done you may paint a green geranium with red leaves and see it as a scarlet geranium with green leaves.

The reason why Shelford Bidwell found grey paper better than white for his disks was partly that the grey cut off the violet, and partly that by reducing the general intensity of the second stimulus, it rendered

[1] *Roy. Soc. Proc.*, vol. 60, p. 368, vol. 61, p. 269, vol. 68, p. 263.

the first more effective, and thus compensated for the greater distance from the light of the coloured objects. By placing these at a greater distance from the disk it is possible to have a separate lamp to illuminate them.

In any case, to get the best results the lights used must be the same as those for which the pictures were painted. To substitute Osram lamps for carbon fila-ments or vice versa may completely spoil the effect.

The following experiments belong to the second class and illuminate what is known as Areal Induction.

Areal Induction.

On the middle of a patch of green some four or five centimetres across is placed a black spot about half a centimetre in diameter. Observing this behind an ordinary Shelford Bidwell disk it is found that the black spot appears bright purple, showing that the retina has been blinded to green in a part on which *no green light has fallen,* and that the effect is due to *'induction' from the surrounding area.*

In this, as in the other cases, it is most instructive to drive the apparatus for a few moments at high speed, and observe how the special phenomena vanish, to reappear as the disks slow down again.

Also, the effect of running them backwards should be noted.

If the black spot is too large, or what amounts to the same thing, if the observer stands too near, the effect of areal induction will *not reach to the middle of it,* and it will appear purple at the edges only. By watching closely it is possible to see that the effect travels from

the edge of the spot towards its centre. This should be
noted before proceeding to the next experiment.·

Benham's Spectrum Top.

This consists of a disk half black and half white, and
on the white part a series of concentric arcs each occupying about 30°, there being from 18 to 24 of them
according to the size of the disk. The outermost
arc begins against the black on one side and the
innermost ends against the black on the other side, the
rest forming a regular series between, each one beginning
a little later and ending a little sooner.

When such a disk is rotated it appears coloured with
a passable imitation of the spectrum, being red at the
periphery, green midway, and blue at the centre, or red
at the centre and blue at the periphery according to
which way it revolves.

For a given thickness of line there is a certain distance at which the colours look brightest. Seen from
a greater distance they appear pale, and if the observer
is too close each line will seem to have a black centre
and to be coloured only on the edges, thus showing
conclusively that the effect is one of areal induction.

Sherrington's Disk.[1]

This is one of a number of disks described by
Sherrington, but it is the one that bears in the simplest
and most direct manner on the subject of areal induction.
It consists of a disk half black and half chrome-yellow,
on which are two rings, each composed half of black

[1] *Journal of Physiology*, vol. xxi, 1897, p. 38.

and half of ultramarine blue. But one ring has the black half on the black of the disc and the blue half on the yellow, and the other has the black half on the yellow and the blue on the black of the disk.

Thus so far as *successive contrast* is concerned both rings are alike—each passes from blue to black and back again from black to blue once in every revolution. But the *areal induction*, i. e. the effect on the surrounding parts of the retina of a sudden light on any one portion, is much greater in the second case. For in the first we have only blue and yellow side by side—the black on black being non-effective. In the second we have yellow against black for half a revolution followed by black against blue for the other half.

The difference is very marked. It requires a higher speed to extinguish ' flickering ', in the ring for which areal induction is greatest.

The rates for both rings should be measured. To eliminate any possible influence of radial distance— i. e. difference of velocity of translation—I employ three rings, one on each side of that which gives the greatest areal induction.

VI. 13. After-effects of Moving Bodies.

If we spin round quickly three times or more, when we stand still all the world appears to be moving slowly in the opposite direction. That is the fundamental experiment of this section. But it will be shown that it is only one manifestation, in its simplest form, of a very remarkable faculty in which more than one sense organ is concerned.

As thus produced, this sensation of vertigo, or familiarly giddiness, is ascribed to the semi-circular canals. While it lasts, it lessens or destroys the power of balancing, and if continued produces headache or even vomiting. But though in its most familiar form it is produced when the whole body undergoes rhythmic movements as on board ship, in its more complex forms the body is not moved at all.

Experiment 1. Throw on the screen a picture of a large wheel which can be made to rotate steadily. After watching it for some time, stop it suddenly. It will appear to revolve slowly in the opposite direction. This is an example of a simple movement apparently reversed.

Experiment 2. Substitute for the wheel a boldly drawn spiral like the hair-spring of a watch. This on rotation will seem to be continually expanding or contracting according to the way it turns. And on stopping it, this *apparent motion* will be reversed.

Experiment 3. Project upon the screen a ' chromatrope'. These, which used to be very popular at lantern exhibitions, consist of a pair of disks pierced with radial or spiral slits and made to rotate in front of one another in opposite directions. They were frequently painted of various colours. The result is that a series of coloured flames appear to shoot out from the centre towards the periphery, and when the movement of the disks is suddenly stopped this apparent movement is reversed, and streaks of light, really stationary, seem to be crowding towards the centre. In the more complex chromatropes there are several concentric rings with the spiral lines sloping alternately one way and the other. In the after-effect *each ring gives rise to an apparent motion contrary to its own.*

The thing to notice is, that whereas giddiness affects the senses generally, this phenomenon affects each part of the retina separately and independently. Moreover, that it is not a mere mental association of the idea of movement is shown by the fact that we can substitute another object. After gazing at the chromatrope, look at a sheet of mottled grey paper—it will seem to copy in the reverse direction the movements of the disk.

When in the train passing through flat open country, fix your eyes for a while on the middle distance, and then look down suddenly at the mat by your feet. It will seem to be the seat of a curious squirming motion— some threads of it travelling rapidly in one direction, others, close to them, moving slowly the opposite way. If you are lucky enough to be in an end carriage with an observation window, fix your eyes on the rails about fifty yards from the car, and keep them so until they begin to feel tired, not gazing from object to object, but keeping steadily to one fixed distance. Then look at your paper. The lines will appear to be coming towards you out of the distance with the middle words growing bigger every moment.

If you are travelling towards the vanishing point instead of away from it, the after-effect shows everything getting smaller as you look at it, instead of bigger.

The rapid flow of water just above a waterfall will often serve to excite these phenomena.

It is evident that the peripheral regions of the eye are very sensitive to movement of the images falling on them. They give very little definition—less indeed than we are apt to imagine—but they call our attention instantly to anything that moves, that we may glance at it.   Probably some such function has been of enormous value to the race in the past—at any rate, we have in these 'after-effects of moving bodies' evidence that the retina *keeps watch, not merely on one object at a time, but on a dozen different movements* going on around the thing we are actually looking at.

Oxford : Horace Hart, M.A., Printer to the University